# Environmental Management in Construction

# Also available from Taylor & Francis

## **Risk Management in Projects**

Martin Loosemore, John Raftery,
Charles Reilly, David Higgon

| | | |
|---|---|---|
| | Taylor & Francis | Hb: 0–415–26055–8 |
| | | Pb: 0–415–26056–6 |

## **Construction Project Management**

Peter Fewings

| | | |
|---|---|---|
| | Taylor & Francis | Hb: 0–415–35905–8 |
| | | Pb: 0–415–35906–6 |

## **Practical Construction Management**

R.H.B. Ranns, E.J.M. Ranns

| | | |
|---|---|---|
| | Taylor & Francis | Pb: 0–415–36257–1 |

## **Human Resource Management in Construction Projects**

Martin Loosemore, Andrew Dainty, Helen Lingard

| | | |
|---|---|---|
| | Spon Press | Hb: 0–415–26163–5 |
| | | Pb: 0–415–26164–3 |

## **Understanding I.T. in Construction**

Rob Howard, Ming Sun

| | | |
|---|---|---|
| | Spon Press | Pb: 0–415–23190–6 |

Information and ordering details

For price availability and ordering visit our website
**www.tandf.co.uk/builtenvironment**
Alternatively our books are available from all good bookshops.

# Environmental Management in Construction

## A quantitative approach

# Zhen Chen and Heng Li

Taylor & Francis
Taylor & Francis Group

LONDON AND NEW YORK

First published 2006
by Taylor & Francis
2 Park Square, Milton Park, Abingdon, Oxon OX14 4RN

Simultaneously published in the USA and Canada
by Taylor & Francis
270 Madison Ave, New York, NY 10016, USA

*Taylor & Francis is an imprint of the*
*Taylor & Francis Group, an informa business*

© 2006 Zhen Chen and Heng Li

Typeset in Times by
Integra Software Services Pvt. Ltd, Pondicherry, India
Printed and bound in Great Britain by
Antony Rowe Ltd, Chippenham, Wiltshire

*British Library Cataloguing in Publication Data*
A catalogue record for this book is available
from the British Library

*Library of Congress Cataloging in Publication Data*
Chen, Zhen, 1967–
    Environmental management in construction : a quantitative approach /
Zhen Chen and Heng Li.
       p. cm.
    Includes bibliographical references and index.
    ISBN 0–415–37055–8 (hardback : alk. paper) 1. Building—Data
processing.    2. Construction industry—Environmental
aspects—Measurement.    3. Environmental protection—
Management.    4. Sustainable buildings—Design and construction
I. Li, Heng.  II. Title.

    TH437.C435 2006
    690.028′6—dc22

                                                    2005031998

ISBN10: 0–415–37055–8 (hbk)
ISBN10: 0–203–03036–2 (ebk)

ISBN13: 978–0–415–37055–4 (hbk)
ISBN13: 978–0–203–03036–3 (ebk)

# Contents

## 1 Introduction

*1.1 Overview 1*

*1.2 Objectives of the book 1*

*1.3 Organization of the book 2*

    *1.3.1 Chapter 2: E+: An integrative methodology 2*

    *1.3.2 Chapter 3: Effective prevention at pre-construction stage 3*

    *1.3.3 Chapter 4: Effective control at construction stage 4*

    *1.3.4 Chapter 5: Effective reduction at post-construction stage 5*

    *1.3.5 Chapter 6: Knowledge-driven evaluation 5*

    *1.3.6 Appendices 6*

## 2 E+: An integrative methodology

*2.1 Introduction 7*

*2.2 Background 8*

*2.3 A questionnaire survey 10*

    *2.3.1 Data collection 10*

    *2.3.2 Overall status 10*

    *2.3.3 Main reasons for indifference 10*

*2.4 Examinations 17*

    *2.4.1 Governmental regulations 17*

    *2.4.2 Technology conditions 18*

# Tables

# Figures

# About the authors

**Zhen Chen** is Research Fellow of the *Innovative Design and Construction for People* Project (part of the programme on *Sustainable Urban Environments* funded by the UK Engineering & Physical Sciences Research Council (EPSRC)) at The University of Reading. He received his B.Sc. degree in building engineering from Qingdao Technological University in Tsingtao, his M.Sc. degree in construction engineering from Tongji University in Shanghai, and his Ph.D. in construction management under **Heng's** supervision from The Hong Kong Polytechnic University. Since 1990, he has been working as an academic at several universities in China, New Zealand, and the UK, including Qingdao Technological University, Tongji University, The Hong Kong Polytechnic University, Massey University, Loughborough University and The University of Reading. He has worldwide professional experience as a freelance consultant to more than 100 projects. He has generated more than 150 publications and consultation reports covering a wide range of topics related to construction engineering and management. His previous books include *Intelligent Methods in Construction* and *Handbook of Building Construction*. He has research interests in knowledge management.

**Heng Li** is Professor in the Department of Building and Real Estate at The Hong Kong Polytechnic University. He started his academic career from Tongji University in 1987. He then researched and lectured at the University of Sydney, James Cook University, and Monash University before joining The Hong Kong Polytechnic University. During this period, he has also worked with engineering design and construction firms and provided consultancy services to both private and government organizations in Australia, Hong Kong, and China. He has conducted many funded research projects related to the innovative application and transfer of construction information technologies, and he has published 2 books, more than one hundred papers in major journals of his field and has presented numerous conference papers in proceedings. His previous books include *Machine Learning of Design Concepts* and *Implementing IT to Obtain a Competitive Advantage in the 21st Century*. He is a review editor of the *International Journal of Automation in Construction* and holds editorships of six other leading journals in his area of expertise. His research interests include intelligent decision-support systems, product and process modelling, and knowledge management.

# Foreword

The interconnectedness between resources consumption and the activities of individuals in environmental management is often appreciated in qualitative regulations, but sometimes it is not sufficiently recognized in quantitative studies. Too frequently the implications of how the interaction between all elements of an environmental management system influence the enterprise, project, or process is left only to descriptive prose. It is only recently that technologies have been developed which enable practitioners to assess potential environmental risks in construction management. These technologies now allow practitioners to conduct environmentally-oriented management with information systems which have knowledge bases embedded within them. This book presents a quantitative approach to environmental management based on an integration of an effective decision-making model with a knowledge re-use framework and a system for quantifying environmental impacts of construction activities for complex environmental management of construction projects. Case studies have been provided to illustrate to practical uses of the quantitative methods presented in the book.

The integrated approach to environmental management presented in this book is a very useful contribution to the development of environmental management systems. It suggests a helpful tool for both academics and practitioners to make progress in avoiding the mistakes of the past and to encourage the promotion of sustainable resource utilization in future construction project management.

Professor Peter Brandon DSc MSc FRICS ASAQS
Director of Strategic Programmes in the School of
Construction and Property Management and
Director of the Salford University "Think Lab"
Vice Chairman, the RICS Research Foundation

# Preface

Strategic environmental management under the ISO 14000 series of environmental management standards requires tactical approaches to support its implementation. For this reason, the authors developed a set of quantitative approaches to minimizing adverse environmental impacts in the construction industry. The primary aim of this book is to demonstrate how quantitative approaches can be made serviceable to environmental management in the construction industry. Specifically, the book illustrates how quantitative methods can be applied to measure the degree of adverse environmental impacts that are generated by construction activities onto the surrounding areas, and how to reduce such impacts through minimizing the wastage of materials and equipments, and maximizing the re-use, recycling, and recovery of construction wastes in the construction industry. In addition to the quantitative approaches, a knowledge-driven system for effective environmental management in construction is also presented.

The uniqueness of this book is reflected in three aspects. First, it has comprehensive coverage of literature related to the field of environmental management in construction. Second, it is the first book that presents an integrated system which can quantitatively control and manage adverse environmental impacts generated from construction activities. Third, it presents a knowledge-driven framework which can be conveniently implemented into a computer-based system to further support effective environmental management in construction.

This book is ideal as a textbook for both undergraduate and postgraduate students in construction engineering and management related fields.

Zhen Chen & Heng Li
2006

# Acknowledgements

The authors would like to acknowledge several publishers, including ASCE, Elsevier B.V., Blackwell Publishing, and Hodder Arnold, for their permissions to re-use some contents of previously published journal papers by the authors themselves in this book. All papers previously published by these publishers are cited in the context and listed in the References of this book. These include the following:

> Chen, Z., Li, H., and Wong, C.T.C. (2005). EnvironalPlanning: an analytic network process model for environmentally conscious construction planning. *Journal of Construction Engineering and Management*, ASCE, 131(1), 92–101.
>
> Chen, Z., and Li, H. (2005). A knowledge-driven management approach to environmental-conscious construction. *International Journal of Construction Innovation*, Hodder Arnold, 5(1), 27–39.
>
> Chen, Z., Li, H., and Hong, J. (2004). An integrative methodology for environmental management in construction. *Automation in Construction*, Elsevier, 13(5), 621–628.
>
> Chen, Z., Li, H., Shen, Q.P., and Xu, W. (2004). An empirical model for decision-making on ISO 14000. *Construction Management and Economics*, 22(1), 55–73.
>
> Chen, Z., Li, H., and Wong, C.T.C. (2003). Webfill before landfill: an e-commerce model for waste exchange in Hong Kong. *Journal of Construction Innovation*, Hodder Arnold, 3(1), 27–43.
>
> Chen, Z., Li, H., and Wong, C.T.C. (2002). An application of bar-code system for reducing construction wastes. *Automation in Construction*, Elsevier, 11(5), 521–533.
>
> Chen, Z., Li, H., and Wong, C.T.C. (2002). Webfill before landfill: an e-commerce model for waste exchange in Hong Kong. *Journal of Construction Innovation*, Hodder Arnold, 3(1), 27–43(17).
>
> Chen, Z., Li, H., and Wong, C.T.C. (2000). Environmental management of urban construction projects in China. *Journal of Construction Engineering and Management*, ASCE, 126(4), 320–324.

Li, H., Chen, Z., and Wong, C.T.C. (2001). Application of barcode technology for an incentive reward program to reduce construction wastes in Hong Kong. *Computer-Aided Civil and Infrastructure Engineering*, Blackwell, 18(4), 313–324.

Li, H., Chen, Z., Wong, C.T.C., and Love, P.E.D. (2002). A quantitative approach to construction pollution control based on resource leveling. *International Journal of Construction Innovation*, Hodder Arnold, 2(2), 71–81.

The authors would also like to acknowledge the contribution of all who have put their efforts in relevant collaborative research and in this book. We would also like to express our thanks to Mr Tony Moore, Senior Editor, Taylor & Francis Books; Dr Monika Faltejskova, Editorial Assistant, Taylor & Francis; Ms Caroline Mallinder, Publisher, Taylor & Francis; and Ms Sunita Jayachandran, Project Manager, Integra Software Services, for their very valuable contributions.

# List of abbreviations

| | |
|---|---|
| A&I | Adoption and Implementation |
| AHP | Analytic Hierarchy Process |
| ANN | Artificial Neural Network |
| ANP | Analytic Network Process |
| C&D | Construction and Demolition |
| CM | Construction Management |
| CPI | Construction Pollution Index |
| EIA | Environmental Impact Assessment |
| EM | Environmental Management |
| EMS | Environmental Management System |
| EPA | External Patch Antenna |
| ERP | Enterprise Resource Planning |
| ESS | Environmental Supervision System |
| E3 | Effective, Efficient, and Economical |
| FIP | Financial Incentive Program |
| GA | Genetic Algorithm |
| GIS | Geographical Information System |
| GPS | Global Positioning System |
| IRP | Incentive Reward Programme |
| IT | Information Technology |
| KB | Knowledge Base |
| KM | Knowledge Management |
| KMS | Knowledge Management System |
| LCA | Life Cycle Assessment |
| M&E | Materials and Equipments |
| PDA | Personal Digital Assistants |
| PERT | Programme Evaluation and Review Technique |
| RC | Reinforced Concrete |
| SPPI | Stochastic Process Pollution Index |
| SWOT | Strengths, Weaknesses, Opportunities, and Threats |
| TTS | Trip-Ticket System |
| VLD | Vehicle Location Device |
| WAN | Wide Area Network |

# Chapter 1

# Introduction

## 1.1 Overview

Adverse environmental impacts of construction such as soil and ground contamination, water pollution, construction and demolition (C&D) waste, noise and vibration, dust, hazardous emissions and odours, demolition of wildlife and natural features and archaeological destruction have been major concerns since early 1970s and received more and more attention in the construction industry, especially after the BS 7750 and the ISO 14001 Environmental Management System (EMS) were promulgated one after another in the 1990s.

However, although there have been many academic studies and professional practices for environmental management (EM) in construction, many of them were conducted in the form of regulations or guidelines. A literature review conducted by the authors of this book from the ASCE's CEDB (Civil Engineering Database) and the EI's Compendex® databases (refer to Table 3.6) revealed that only 2% of works provide quantitative methods in the total number of publications related to EM in construction in 2003. In this book, a set of quantitative methods, which finally composes an integrative prototype for supporting the EM in construction, is presented to support the EM in the lifecycle of a construction project.

## 1.2 Objectives of the book

The objective of this book is to describe an integrative quantitative approach to EM in construction. This objective has been achieved through five steps. First of all, an integrative methodology named E+ for dynamic environmental impact assessment (EIA) in construction is developed as a comprehensive framework. Next, four analytical methods are developed and integrated. These four methods include the construction pollution index (CPI) method to quantitatively evaluate and reduce pollution and hazard levels of processes and projects, the env.Plan method to evaluate environmental-consciousness of proposed construction plans and select the prime environmental-friendly construction plan, the incentive reward program (IRP) method to reduce on-site construction wastes through an incentive reward programme, and the Webfill method to promote C&D waste exchange in the local construction industry. Finally, the implementation of the integrative analytical approach is demonstrated by an experimental case study.

## 1.3 Organization of the book

There are eight chapters in this book. These chapters are organized according to their relationships with the objectives of the book. To start with the introduction to the integrative prototype for EM in construction, the need for quantitative approach to EM in construction is presented based on previous investigations on adoption and implementation of ISO 14001 EMS in construction enterprises in Australia, Hong Kong, mainland China, Singapore, United Kingdom and United States, etc. After the integrative prototype (named E+) for dynamic EIA in construction is described in Chapter 2, four practical analytical methods – including CPI method, env.Plan method, IRP method, and Webfill method, together with their working knowledge bases (KBs), which are essential components in the E+ prototype – are elaborated individually from Chapters 3 to 5. For the application of the E+ prototype to EM in construction, an experimental case study is then conducted to demonstrate the developed E+ prototype in Chapter 6. In addition to the E+ prototype and its essential components, conclusions and recommendations are then presented in Chapter 6 to summarize contributions and limitations of this book, and recommend further research and development for quantitative EM in construction. Finally, four appendices have been provided: a questionnaire for an investigation on the acceptability of the ISO 14001 EMS in the construction industry, a decision-making model for acceptance of the ISO 14001 EMS, sample waste exchange websites, and the function menu of Webfill (an e-commerce business plan). The abstract of each chapter is as follows.

### 1.3.1 Chapter 2: E+: An integrative methodology

The ISO 14001 EMS is not as widely acceptable as the EIA process in the construction industry, according to previous investigations. In order to demonstrate the acceptability of the ISO 14001 EMS in the construction industry, this chapter reports a remarkable disagreement between the rate of the ISO 14001 EMS registration and the rate of implementation of EIA in the Chinese construction industry. This disagreement indicates that the contractors there might not have really applied EM in construction projects. This hypothesis is then examined in this chapter by a questionnaire survey conducted among 72 main contractors in Shanghai, mainland China. Survey results indicate that there are five classes of factors influencing the acceptability of the ISO 14001 EMS, including governmental laws and regulations, technology conditions, competitive pressures, cooperation attitude, and cost–benefit efficiency. Reasons why approximately 81% of contractors surveyed are indifferent to the ISO 14001 EMS are then analysed based on the critical classes. A linear discriminant model for decision-making on whether to accept the ISO 14001 EMS for construction companies is consequently developed and provided in Appendix B.

On the other hand, the remarkable difference between the registration rate of ISO 14001 EMS and the implementation rate of EIA in the construction industry in mainland China also indicates that there may be little coordination between the implementation of EIA and EMS in construction projects in mainland China, and the EIA practice may not really serve as a tool to promote EM in construction. Since the China Environmental Protection Bureau enacted laws to implement the environmental supervision system in construction project supervision, contractors have to pay greater attention to adopt and implement EM in construction. According to the second emphatic factor based on the survey results, contractors paid greater attention to technology conditions on both construction and management and they thought the technology conditions can effectively enhance their working efficiency in EM in construction. Based on this consideration, this chapter presents an integrative methodology named E+ for dynamic EIA in construction, which integrates various EM approaches with a general EMS process throughout all construction stages in a construction project. As the E+ is designed to be a general tool to conduct EM in construction, it is expected to assist contractors to effectively, efficiently, and economically enhance their environmental performances all over the world.

### 1.3.2 Chapter 3: Effective prevention at pre-construction stage

To the authors' knowledge, there have been very few studies on integrating concerns of EM in the construction planning stage in particular. Construction planning involves the choice of construction technology, equipment and materials, the definition of work tasks, the layout of construction site, the estimation of required resources and durations for individual tasks, the estimation of costs, the preparation of a project schedule, and the identification of any interactions among the different work tasks, etc. (Horvath and Hendrickson 1998; Hendrickson and Horvath 2000). As a fundamental and challenging task, construction planning should not only strive to meet common concerns such as time, cost, and quality requirement, but also explore possible measures to minimize environmental impacts of the projects at the outset.

From this point of view, this chapter presents two quantitative methods for EM at pre-construction stage: the CPI method to quantitatively evaluate and reduce pollution and hazard levels of construction processes and projects, and the env.Plan method to quantitatively evaluate environmental-consciousness of proposed construction plan alternatives and thereafter select the prime environmental-friendly construction plan. Both CPI method and env.Plan method can greatly facilitate the application of the E+ prototype at the pre-construction stage.

The CPI method is a quantitative approach to EM on pollution and hazards potentially caused by construction projects in accordance with a proposed construction plan. The proposed CPI method is to assess and control the potential environmental problems upon implementation of a construction plan, and a method to calculate the CPI is originally put forward which provides a quantitative measurement of pollution and hazards caused by construction projects. In addition to the conception of the CPI, a practical method to comprehensively reducing construction pollution level during construction is put forward and examined. The CPI method is further applied in a commercial software environment, i.e. Microsoft Project©. A comparison study on the performance of CPI levelling between the normally used resource levelling method and genetic algorithm (GA) is also conducted. The parameters of CPI, i.e. pollution and hazards magnitude $(h_i)$, are treated as a pseudo resource and integrated with a construction schedule. When the level of pollution for site operations exceeds the permissible limit identified by a regulatory body, the GA-enhanced levelling technique is used to reschedule project activities so that the level of pollution can be re-distributed and thus reduced. The GA-enhanced resource levelling technique is demonstrated using 20 on-site construction activities in a project. Experimental results indicate that the GA-enhanced resource levelling method performs better than the traditional resource levelling method used in Microsoft Project©. The proposed method is an effective tool that can be used by project managers to reduce the level of pollution at a particular period of time, when other control methods fail. The CPI is a primary component of the E+ prototype for reducing potential adverse environmental impacts during construction planning stage.

Although the CPI method is an effective and efficient approach to reducing or mitigating pollution level during the construction planning stage, the problem of how to select the best construction plan based on distinguishing the degree of its potential adverse environmental impacts is still unsolved. In the second section of this chapter, the authors review essential environmental issues and their characteristics in construction, which are critical factors in evaluating potential adverse environmental impacts of a construction plan. These environmental indicators are then chosen to structure two decision models for environmental-conscious construction planning by using an analytic network process (ANP), including a complicated model and a simplified model. The two ANP models named env.Plan can be applied to evaluate potential adverse environmental impacts of alternative construction plans. The env.Plan method is an important component of E+ prototype in selecting most environmental-friendly construction plan alternatives, and it is also a necessary complement of the CPI method in the E+ prototype.

### 1.3.3  Chapter 4: Effective control at construction stage

This chapter presents a group-based IRP method to encourage site workers to minimize avoidable wastes of construction materials by rewarding them according to the amounts and values of materials they saved. Based on the formulations of the IRP, bar-code technique is used to facilitate effective, efficient, and economical management of construction materials on site. In addition to the integration of the group-based IRP and the bar-code technique for reducing construction waste, an IRP-integrated construction management (CM) system is also introduced to avoid jerry-building and solve rescheduling problems due to rework because of quality failure. For the application of the IRP method, an experimental research is then conducted on a residential project in Hong Kong. Results from the experimental research demonstrate the effectiveness of the IRP in motivating workers to reduce construction wastes. In addition to the IRP method and its implementation, discussions on the relationship between construction waste reduction and time-cost performances, and difficulties and challenges of applying the IRP method are presented accordingly. The IRP method is a basic component of E+ prototype used for minimizing avoidable material wastes on construction site.

### 1.3.4 Chapter 5: Effective reduction at post-construction stage

Although the trip-ticket system (TTS) has been widely implemented to manage C&D waste in many countries for a long time, problems still exist in the landfill disposal of C&D waste. For example, it is reported that fees are difficult to collect from waste transporters for tipping the C&D waste at the landfill site in Hong Kong. Based on an examination on the flexibility of currently enacted TTS for reducing C&D waste, this chapter proposes an e-commerce model named Webfill in order to facilitate traditional TTS to effectively, efficiently, and economically manage C&D waste in macro scopes of the construction industry. The computational structure of the Webfill system is therefore described and the usefulness of the Webfill method is accordingly evaluated based on computer simulations which provide a direct comparison between the existing TTS and the Webfill-enhanced TTS. The Webfill method is an enhanced component of E+ prototype for reducing, reusing, and recycling C&D waste inside and outside a construction enterprise at post-construction stage when C&D waste has been inexorably generated.

### 1.3.5 Chapter 6: Knowledge-driven evaluation

This chapter demonstrates an integrative application of the E+ prototype for dynamic EIA in construction illustrated in Chapter 2 by using an experimental case study, in which various quantitative EM methods described in Chapters 3–5 are integrated with a general ISO 14001 EMS process throughout all construction stages in a construction project. Besides the demonstration of the E+ prototype, the experimental case study used in this chapter also indicates that it is necessary to further develop the integrative prototype to be a Web-based E+

environment to effectively, efficiently, and economically undertake and enhance EM in construction.

### 1.3.6 Appendices

The appendix section consists of four appendices: Appendix A: a questionnaire for investigating the acceptability of the ISO 14001 EMS in the construction industry, Appendix B: a decision-making model for acceptance of the ISO 14001 EMS, Appendix C: sample waste exchange websites, and Appendix D: the function menu of Webfill (an e-commerce business plan). Appendices A and B complement the investigation on the acceptability of ISO 14001 EMS in the construction industry with a questionnaire and corresponding statistic analysis. Appendix C provides a list of 36 websites related to C&D waste exchange from which the e-commerce model for the Webfill method is developed. Appendix D illustrates the function menu of Webfill (an e-commerce business plan).

# Chapter 2

# E+: An integrative methodology

## 2.1 Introduction

Since September 1996, when the ISO 14000 series was first issued, environmental management systems (EMSs) have been received in the construction industry globally (ISO 2001), and have become a research and development area in construction management (Kein *et al.* 1999; Ofori *et al.* 2000; Tse 2001). The ISO survey in 2001 showed that there is a continuing strong growth of ISO 14001 EMS registration in the construction industry; for instance, the number of registered companies increased from 298 as at the end of 1998, to 500 as at the end of 1999, and then up to 1035 as at the end of 2000 (ISO 2001). However, three statistical figures from mainland China indicate that the EMS has not been prevalent in the construction industry there. The first figure is the percentage of environmental certificates awarded to Chinese enterprises versus total environmental certificates awarded to enterprises worldwide, which is as low as 2.23% (ISO 2001); the second figure is the percentage of environmental certificates awarded to Chinese construction enterprises versus total environmental certificates awarded to Chinese enterprises, which is as low as 7.65% (ISO 2001); and the third figure is the percentage of the construction enterprises that have been awarded environmental certificates versus total governmental registered construction enterprises in mainland China, which is as low as 0.083% (CCEMS 2001; CEC 2001; CEIN 2001a; CACEB 2002). These statistical data indicate that the construction enterprises have not fully accepted the ISO 14000 series in mainland China.

By contrast, a higher implementation rate of environmental impact assessment (EIA) in construction projects in mainland China is encountered from another statistical analysis (China EPB 2000/2001). The EIA of construction projects is the process or technique of identifying, predicting, evaluating, and mitigating the biophysical, social, and other relevant environmental effects of development proposals or projects prior to major decisions being taken and commitments made (IAIA 1997; European Commission 1999; landscape Institute with IEMA 2002). According to the *Official Report on the State of the Environment in China 2000* (China EPB 2000/2001), the implementation rates of EIA were 90.4% in 1999 and 94.8% in 2000. A further investigation on the implementation rate of

EIA in mainland China indicates that the average EIA rate from 1995 to 1997 is 82% (a mean of three yearly average EIA rates, which are 79% in 1995, 81% in 1996, and 85% in 1997). Comparing with what it was in 1999 and 2000, the implementation rate of EIA is rising, although it varies in different municipalities and provinces in a range from 46 to 100%. It is obvious that the EIA rate is much higher than the implementation rate of the ISO 14000 series in mainland China.

The statistical data indicates that the ISO 14000 series have not yet been widely implemented in the Chinese construction industry and the problem of whether contractors have really accepted the standard also emerges. In order to further verify the observation and understand the reasons that hinder the acceptance of the standard, a questionnaire survey focusing on the adoption and implementation (A&I) of EMS and the ISO 14000 series has been conducted over 100 selected construction companies in Shanghai, which is selected as a representative city in mainland China. Reasons why some contractors surveyed resist the A&I of the ISO 14000 series ($ISO\ 14Ks_{A\&I}$) are then analysed and useful conclusions, including a discriminant model for decision-making on ISO 14000 acceptance, are generated. A Microsoft Excel® spreadsheet is adopted to apply the discriminant model.

## 2.2   Background

Environmental management in construction has received more and more attention since the early 1970s. For example, studies on noise pollution (U.S.EPA 1971), air pollution (Jones 1973), and solid waste pollution (Skoyles and Hussey 1974; Spivey 1974a,b) from construction sites were individually conducted in the early 1970s. Although the expression "EM in construction" came out in the early 1970s after the U.S. National Environmental Policy Act of 1969 was enacted (Warren 1973), the concept of EM in construction was introduced in the late 1970s, when the role of environmental inspector was defined in the design and construction phases of projects to provide advice to construction engineers on all matters in EM (Spivey 1974a,b; Henningson 1978). However, there had been little enthusiasm for establishing an EMS in construction organizations until two important standards, BS 7750 (issued in 1992) and the ISO 14000 series (issued in 1996), were promulgated to guide the construction industry from passive construction management on pollution reduction to active EMS for pollution prevention.

In the 1990s, the Construction Industry Research and Information Association (CIRIA) conducted a series of reviews on environmental issues and have under-taken initiatives relevant to the construction industry after the introduction of BS 7750 (Shorrock *et al.* 1993; CIRIA 1993, 1994a,b, 1995; Guthrie and Mallett 1995; Petts 1996). Thereafter, research efforts for EM have also been put into the implementation of EMS and the accreditation of ISO 14001 EMS by authoritative institutions in the construction industry, including the CIOB (Clough and Antonio 1996), the FIDIC (1998), the Construction Policy Steering Committee (CPSC 1998), and the CIRIA (Uren and Griffiths 2000).

In order to assess the extent of EMS implementation within the construction industry, several investigations have been conducted. For example, Kein *et al.* (1999) conducted a field study in Singapore to assess the level of commitment of ISO 9000-certified construction enterprises to EM. They found that contractors in Singapore were aware of the merits of EM, but were not instituting systems towards achieving it; Ofori *et al.* (2000), also in Singapore, then conducted a survey to ascertain the perceptions of construction enterprises on the impact of the implementation of the ISO 14000 series on their operations. Major problems were identified, such as the shortage of qualified personnel, lack of knowledge of the ISO 14000 series, indistinct cost–benefit ratio, disruption and high expenses on changing traditional practices, and resistance from employees, etc.; the CIRIA (1999) led a self-completion questionnaire survey of the state of environmental initiatives within the construction industry and of sustainability indicators for the civil engineering industry in the United Kingdom; Tse (2001) conducted an independent questionnaire survey in the Hong Kong construction industry to gain a further understanding of the difficulties in implementing the ISO 14000 series; Lo (2001), also in Hong Kong, made an effort to identify nine critical factors for the implementation of ISO 14001 EMS in the construction industry based on critical factors drawn from an investigation in another industry; and the CPSC (2001), in Australia, conducted a questionnaire survey of the New South Wales construction industry on EM with industry leaders. All these questionnaire surveys aimed to clarify the real situations in $ISO\ 14Ks_{A\&I}$ in local construction industries.

One important contribution of these surveys is that researchers have gained useful insights into the problems and difficulties of implementing the ISO 14000 series. Their survey results provide useful information not only for improving efficiency on EMS implementation but also for developing the EMS itself, focusing on effective EM in the construction industry. For example, Tse (2001) has found four major obstacles in implementing the ISO 14000 series in Hong Kong's construction industry, including lack of government pressure, lack of client requirement or supports, expensive implementation cost, and difficulties in managing the EMS with the current sub-contracting system. One cannot easily draw such constructive conclusions in detail without such a kind of survey. However, what originally impelled us to an investigation on the acceptability of the ISO 14000 series in mainland China was not the advantage of a survey even though there is little published research work in this area, but the remarkable disagreement between the rate of ISO 14001 EMS registration and the rate of EIA implementation in Chinese construction industry. As stated previously, the remarkable deviation between the two rates indicates that the contractors in mainland China may not have really applied EM in construction projects. In order to verify this hypothesis, a questionnaire survey was conducted and details of the questionnaire survey are described below.

## 2.3   A questionnaire survey

### 2.3.1   Data collection

The methodology adopted for this study involves the use of a structured questionnaire (see Appendix A) and a statistical analysis. Shanghai was selected as a representative city. As one of the most industrialized Chinese cities, Shanghai is halfway along the eastern coastline of mainland China. It is a municipality with an urban population of 9.6 million, and plays an essential role in national socio-economic affairs; furthermore, Shanghai is one of the areas where there have been large numbers of construction projects in mainland China in the past several years (China NBS 2000).

In mainland China, construction enterprises are divided into three types: main contractors, specialized contractors, and labour contractors (MOC 2001a,b,c). Each type is further divided into different classes according to the characteristics of construction projects and technological demands. And each class is then divided into different grades with specified qualifications to individual companies. At present, there are five grades of main contractors. They are Special Grade, and Grade-1 to Grade-4. The population of the survey group consists of 100 main building contractors randomly selected from Shanghai, including 50 Grade-1 qualified contractors and 50 Grade-2 qualified contractors.

Hundred copies of the questionnaire were distributed to the main contractors in Shanghai, with whom the authors were acquainted in April 2001. By the end of October 2001, 72 usable responses were received. This represents 1.5% of contractors in the Shanghai construction industry. All survey data accumulated were analysed using a standard version of SPSS® 11.

### 2.3.2   Overall status

Among these 72 construction companies, 2 companies have ISO 14001 EMS registrations, 1 company is under assessment for registration, 11 companies are willing to apply for registration, and 58 companies do not want to apply. These results indicate that the ISO 14000 series has only been accepted by 19% of the contractors surveyed, while others (81%) gave out their indifference to the ISO 14000 series.

### 2.3.3   Main reasons for indifference

The reasons for indifference to the ISO 14000 series are summarized in Table 2.1. The acceptability of the ISO 14000 series is examined separately in terms of A&I in the questionnaire survey (see Parts 6 and 7 in Appendix A), as adoption

Table 2.1 Potential influential reasons for indifference to the ISO 14000 series

(a) Reasons for not adopting the ISO 14000 series

| Class | Reason Item | Grade Mean | Grade Rank |
|---|---|---|---|
| 1 | Lack of governmental administrative requirement on adopting the ISO 14000 series | 9.0 | 1 |
| | Lack of governmental encouragement on financial subsidies, e.g. tax deduction/return | 8.5 | 2 |
| | Lack of governmental encouragement on non-financial allowance | 8.4 | 3 |
| 2 | Lack of reliable consultant companies on tutorship of adoption of the ISO 14000 series | 7.5 | 6 |
| 3 | Lack of competitive pressure from domestic construction industry | 7.1 | 7 |
| | Lack of competitive pressure from international construction industry within WTO | 7.0 | 8 |
| 4 | Lack of internal initiative consciousness on implementation of EMS | 8.0 | 4 |
| 5 | High cost of implementation of ISO 14001 EMS (About RMB 0.3M) | 7.6 | 5 |
| | High cost of ISO 14001 EMS assessment, certification, and maintenance | 6.8 | 9 |
| | Additional cost of human resource on adopting and implementing the ISO 14000 series | 6.8 | 9 |
| | High cost of ISO 14001 registration (About RMB 50,000) | 6.6 | 10 |
| – | Additional cost of reorganization on adopting and implementing the ISO 14000 series | 6.3 | 11 |
| – | The necessity of management involvement on adopting the ISO 14000 series | 6.3 | 11 |
| – | Interrupt and adjustment of construction processes on implementing the ISO 14000 series | 6.1 | 12 |
| – | Entire employees' training and education before implementing ISO 14001 EMS | 6.0 | 13 |
| – | Various additional EM documents on adopting ISO 14000 series | 6.0 | 13 |
| – | Lack of requirement and pressure from clients or suppliers | 6.0 | 13 |
| – | Lack of expectation from clients or suppliers | 6.0 | 13 |
| – | Additional cost on training functionaries inside company | 5.9 | 14 |
| – | Lack of intention to establish enterprise's internal ISO 14000 based EMS | 5.6 | 15 |
| – | Less encouraging subcontractors to adopt ISO 14000 series for improving EM | 5.6 | 15 |
| – | Additional cost of failure on adopting ISO 14001 EMS | 5.2 | 16 |

Table 2.1 (Continued)

(b) Reasons for not implementing the ISO 14000 series

| Class | Item | Grade Mean | Grade Rank |
|---|---|---|---|
| 1 | Lack of pressure from the government | 8.0 | 4 |
| 2 | Multifarious documental operation process of the ISO 14000 series | 9.0 | 2 |
| | Destitute of applicability of the ISO 14000 series in construction enterprises | 8.5 | 3 |
| | Lack of suitable technology and material for environmental protection | 8.0 | 4 |
| 3 | Lack of pressure from the competitors inside construction industry | 6.5 | 6 |
| | No competitors implemented the ISO 14000 series first inside construction industry | 6.0 | 7 |
| | Lack of pressure from the clients | 5.5 | 8 |
| 4 | Lack of correspondence and cooperation of design and construction | 9.0 | 2 |
| | Poor employees' attitude towards cooperation on implementing the ISO 14000 series | 9.0 | 2 |
| | Poor administrators' attitude towards cooperation on implementing the ISO 14000 series | 9.0 | 2 |
| | Poor subcontractors' attitude towards cooperation on implementing the ISO 14000 series | 9.0 | 2 |
| | Poor suppliers' attitude towards cooperation on implementing the ISO 14000 series | 7.5 | 5 |
| 5 | Additional cost of implementation of ISO 14001 EMS | 9.5 | 1 |
| | Impacts and additional expense of construction on interruption and adjustment | 9.5 | 1 |
| | Costly expense on implementation | 9.5 | 1 |
| – | Success/failure on employees' training and education inside enterprise | 8.0 | 4 |
| – | Success/failure on maintenance and continuous assessment of the ISO 14000 series | 8.0 | 4 |
| – | Success/failure on administrator's training and education inside enterprise | 8.0 | 4 |
| – | Success/failure on combination with other EMS inside enterprise | 6.5 | 4 |
| – | Success/failure on adjustment of organizational structure inside enterprise | 6.0 | 7 |

Notes
Class 1 = Governmental regulations; Class 2 = Technology conditions; Class 3 = Competitive pressures; Class 4 = Cooperative attitude; Class 5 = Cost–benefit efficiency.

means only to get an ISO 14001 EMS registration, while the implementation is to carry out the EMS after registration, and some contractors who gain ISO 14001 certificates might not carry out a qualified EMS up to the requirements of the ISO 14000 series. Table 2.1a gives reasons for indifference to adopting the ISO 14000 series, and Table 2.1b gives reasons for indifference to implementing the ISO 14000 series.

In order to find critical factors that influence the adoption and the implementation of the ISO 14000 series, reasons in Table 2.1a and Table 2.1b are assorted into classes according to their coherence, and five classes are identified: governmental command-and-control regulations on *ISO* $14Ks_{A\&I}$ (governmental regulations), applied environmental-friendly technology conditions in construction and management (technology conditions), competitive pressures from both domestic and foreign trades (competitive pressures), attitude towards cooperation with an EM-seeking enterprise on *ISO* $14Ks_{A\&I}$ (cooperative attitude), and cost–benefit efficiency on *ISO* $14Ks_{A\&I}$ (cost–benefit efficiency). All items are ranked according to their mean score grades, which are calculated with corresponding scores from respondents who are indifferent to the adoption of the ISO 14000 series. The average grades of each of the five classes are then determined by using grade means of each corresponding reason in the class.

First, the main reasons for indifference to adopting the ISO 14000 series (refer to Table 2.1a) show that those respondents score highly in a sequence on governmental regulations (Ranks 1 to 3 with an average grade of 8.6), cooperative attitude (Rank 4 with an average grade of 8.0), technology conditions (Rank 6 with an average grade of 7.5), competitive pressures (Ranks 7 and 8 with an average grade of 7.1), and cost–benefit efficiency (Ranks 5, 9, and 10 with an average grade of 7.0).

In terms of indifference to implementation of ISO 14000 series, major reasons in the classes (as shown in Table 2.1b) were identified, which include the cost–benefit efficiency (Rank 1 with an average grade of 9.5), cooperative attitude (Ranks 2 and 5 with an average grade of 9.3), technology conditions (Ranks 2, 3, and 4 with an average grade of 8.6), governmental regulations (Rank 4 with an average grade of 8.0), and competitive pressures (Ranks 6, 7, and 8 with an average grade of 6.0).

Combining the results of Tables 2.1a and 2.1b, the histograms which indicate the opinions of companies surveyed for not adopting and implementing the ISO 14000 series, as shown in Figure 2.1, were obtained.

Additionally, in order to test whether a mean grade differs from a given hypothesized test value in the corresponding column in Tables 2.1a and 2.1b, the one-sample $t$ test method is employed in every calculation on an individual potential influential factor. At the 95% confidence level, the critical value of $t$ with 57 degrees of freedom (i.e. $n - 1 = 58 - 1$) is 2.11. Therefore, as the absolute value of $t$ (here $t = 0$) is less than $+2.11$, it is concluded that the null hypothesis (mean grade) could not be rejected.

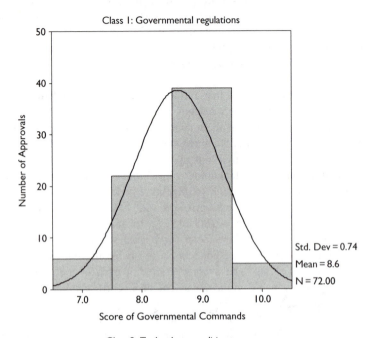

Class 1: Governmental regulations

Std. Dev = 0.74
Mean = 8.6
N = 72.00

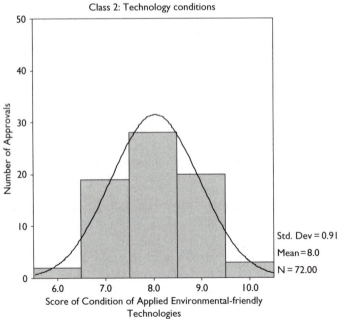

Class 2: Technology conditions

Std. Dev = 0.91
Mean = 8.0
N = 72.00

*Figure 2.1* Class histograms for ISO 14000's acceptability with total 72 respondents.

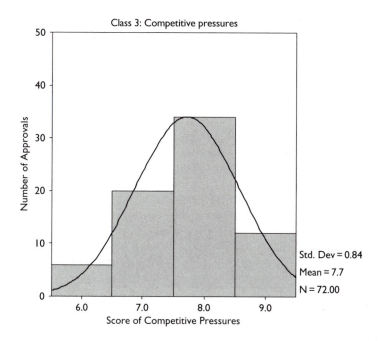

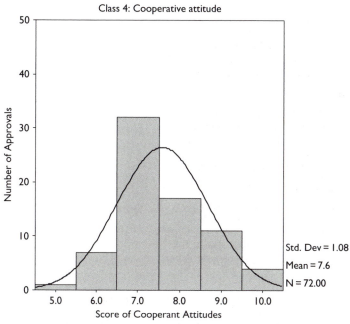

*Figure 2.1* (Continued).

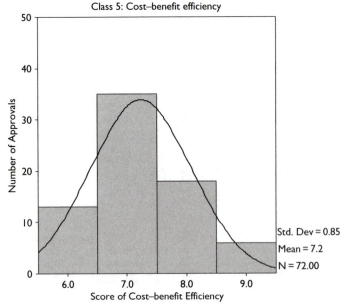

Class 5: Cost–benefit efficiency

Std. Dev = 0.85
Mean = 7.2
N = 72.00

Ranks of top five classes

Ranking with 58 indifferentists

1 Governmental regulations (Mean grade is 8.4)
2 Technology conditions(Mean grade is 8.2)
3 Cooperative attitude (Mean grade is 7.9)
4 Cost–benefit efficiency (Mean grade is 7.4)
5 Competitive pressures (Mean grade is 7.3)

Ranking with 14 accepters

1 Governmental regulations (Mean grade is 8.5)
2 Technology conditions(Mean grade is 7.9)
3 Competitive pressures (Mean grade is 7.8)
4 Cooperative attitude (Mean grade is 7.4)
5 Cost–benefit efficiency (Mean grade is 7.1)

Ranking with 72 respondents

1 Governmental regulations (Mean grade is 8.6)
2 Technology conditions(Mean grade is 8.0)
3 Competitive pressures (Mean grade is 7.7)
4 Cooperative attitude (Mean grade is 7.6)
5 Cost–benefit efficiency (Mean grade is 7.2)

Notes
1 Alpha is a reliability coefficient for rejecting the null hypothesis when in fact the null hypothesis is true.
2 Reliability coefficients of the top five classes: $\alpha = 0.69$, Standardized item $\alpha = 0.70$.

*Figure 2.1* (Continued).

## 2.4   Examinations

According to the survey results, the critical factors for not adopting and implementing the ISO 14000 series are characterized by five classes: governmental regulations, technology conditions, competitive pressures, cooperation attitude, and cost–benefit efficiency. These critical factors are now further analysed.

### 2.4.1   Governmental regulations

The governmental regulations include all kinds of governmental command-and-control ordinances and regulations on encouraging contractors to adopt and implement EMS. In the survey, the governmental regulations are divided into three scopes: administrative requirement on adopting and implementing EMS in construction industry, encouragement of financial subsidies (e.g. tax deduction or repay), and encouragement of non-financial allowance. Analysing data regarding these three kinds of governmental regulations shows that all Pearson's correlation coefficients (0.890 between administrative requirement and financial encouragement, 0.420 between financial and non-financial encouragement, and 0.399 between administrative requirement and non-financial encouragement) are significant at the 0.01 level (2-tailed). Frequencies of each kind of governmental regulation above their mean grades are 76.2, 76.2, and 80.0%; and these frequencies are quite similar on approaching 80%. Moreover, a trend analysis between the governmental regulations and the ISO 14000 series' acceptability indicates that contractors who give higher score to governmental regulations would have less intention to accept the ISO 14000 series. The survey results indicate that the government plays an important role in promoting $ISO\ 14Ks_{A\&I}$, and contractors would prefer to be indifferent to the ISO 14000 series if there were insufficient governmental command-and-control regulations on it.

The survey results offer a conclusion similar to those of the three previous surveys on $ISO\ 14Ks_{A\&I}$ in the construction industry in Hong Kong (Tse 2001) and Singapore (Kein *et al.* 1999; Ofori *et al.* 2000) in that contractors would ignore to adopt and implement the ISO 14000 series directly if there were lack of pressure from the government. The effect of governmental regulations is also reflected in the fact that the high implementation rate of EIA in mainland China is because the *Managerial Ordinance on Environmental Protection of Construction Project* (SC of China 1998) stipulates that all new construction projects must apply for environmental impact approval following an approval procedure of EIA report/form or Ei form before construction. More than 90% of new construction projects have been undertaken according to the EIA procedure and received approval annually in mainland China since the ordinance was issued (China EPB 2001). Moreover, a literature review shows that the governmental regulations particularly affect the number of ISO 14001 certified contractors in Hong Kong.

In the past four years, the number of ISO 14001–certified contractors in Hong Kong was 4 in 1998, 7 in 1999, 4 in 2000, 22 in 2001, and 2 in early 2002 (HKEPD 2002). These numbers coincide with the governmental regulations on promoting the ISO 14000 series issued twice, in later 1996 and early 2000 (HKPC 1996, 2000); for example, there were 15 ISO 14001–certified contractors after the first promotion in 1996 and the figure increased to 39 owing to the second promotion in 2000.

Unfortunately, there have been no governmental regulations on promoting the ISO 14000 series nationally or locally in the Chinese construction industry since 1996, and contractors with less consciousness on environmental protection in mainland China can thus be indifferent to the EMS without any liability. For example, although the Environmental Protection Bureau of China has established seven National Demonstration Districts to display the benefits of implementing ISO 14001 EMS since 1998 (China EPB 2002), there has been no ISO 14000 series–related requirement or restriction for contractors to tender projects (China EPB 2001). Moreover, in the 10th five-year plan of the Ministry of Construction in China (CMC 2000), no environmental-friendly construction technology is promoted. It is thus not surprising to see that near by 81% of contractors were indifferent to the $ISO\ 14Ks_{A\&I}$ in the survey.

### 2.4.2  Technology conditions

Technology conditions refer to the level of environment-friendly or resource-efficient (NAHB Research Center 1999) technologies for reducing negative environmental effects in construction. In the survey, these technologies are divided into three types, the first type includes the use of technologies in order to get accreditation of ISO, the second type includes technologies used for implementing the ISO 14000 series (Technology B), and the third type includes technologies used by a company to reduce negative environmental impacts, although the company does not accept the ISO 14000 series (Technology C). Analysing data regarding these three types of technologies shows that all the Pearson's correlation coefficients (0.469 between Technology A and Technology B, 0.449 between Technology A and Technology C, and 0.442 between Technology B and Technology C) are significant at the 0.01 level (2-tailed). Frequencies of the three types of technologies above mean grades are 76.1, 66.7, and 57.1%, all of which are above 50%. Moreover, a trend analysis between the technology condition and the ISO 14000 series' acceptability indicates that contractors who gave higher scores to the technology condition would be more likely to accept the ISO 14000 series. The survey results indicate that technologies are an important means for adopting and implementing the ISO 14000 series and contractors would prefer to accept the ISO 14000 series if there were sufficient technologies to help them to control and reduce the negative environmental impacts in construction.

### 2.4.3  Competitive pressures

Competitive pressures include pressures from the competitors of both the domestic and international markets on *ISO* 14$Ks_{A\&I}$. The survey divides the competitive pressures into two scopes: domestic competitive pressure and international competitive pressure. Analysing data regarding these two scopes of competitive pressures shows that the Pearson's correlation coefficient (0.558 between domestic competitive pressure and foreign competitive pressure) is significant at the 0.01 level (2-tailed). Frequencies of the two scopes of competitive pressures above mean grades are 64.3 and 61.9%, which are above 60%. Moreover, a trend analysis between the competitive pressures and the acceptability of the ISO 14000 series indicates that contractors who give higher score to the competitive pressures would be more likely to accept the ISO 14000 series. The survey results indicate that competitive pressure is an important consideration when contractors decide whether to adopt and implement the ISO 14000 series, and contractors will accept the ISO 14000 series if there are sufficient competitive pressures.

In the past five years, construction companies in mainland China met with increasing competition from foreign construction companies in the domestic market. According to the statistical data from the China National Bureau of Statistics, the proportion of foreign construction companies has grown with an average rate of 10.7% since 1996, while the proportion of domestic construction companies has shrunk with the rate of 2.9% (China NBS 1998/2000). This indicates that contractors in mainland China are facing severe competition from their international counterparts, especially in the next five to ten years after China's accession to WTO and many important civil infrastructure projects will be tendered internationally (CEIN 19/03/2001).

Unfortunately, contractors involved in the survey have not yet realized the competitive pressure and the trend of globalization, as most of them have been largely accustomed to focusing on competition with their domestic peers.

### 2.4.4  Cooperative attitude

Cooperative attitude reflects the willingness of people in *ISO* 14$Ks_{A\&I}$. In the survey, the cooperative attitude is divided into four scopes: cooperative attitude from designers, cooperative attitude from workers, cooperative attitude from administrators, and cooperative attitude from subcontractors. Analysing data regarding these four scopes of attitude on cooperation shows that the Pearson's correlation coefficients (0.803 for cooperative attitude among workers, administrators, and subcontractors, 0.661 for cooperative attitude between employees and designers, and 0.557 for cooperative attitude among designers, administrators, and subcontractors) are significant at the 0.01 level (2-tailed). Frequencies of the four scopes of attitude on cooperation above mean grades are 59.5, 50.0, 52.4, and 52.4%, all of which are above 50%. Moreover, a trend analysis between the cooperative attitude and the acceptability of the ISO 14000 series indicates that contractors who give higher score to the cooperative attitude would have greater intention of accepting the ISO 14000 series. The survey results indicate that the cooperative attitude towards *ISO* 14$Ks_{A\&I}$ also affects the progression of EMS, and contractors would have accepted the ISO 14000 series if there had been satisfactory cooperation on EMS both inside and outside their companies.

### 2.4.5   Cost–benefit efficiency

Cost–benefit efficiency includes all concerns regarding benign cost–benefit circulations on *ISO* 14$Ks_{A\&I}$ inside a construction enterprise. In our survey, the concerns of cost–benefit efficiency are divided into three main scopes: costs for registration and maintenance of ISO 14001 EMS certification, costs for implementation of ISO 14001 EMS, and benefits from the *ISO* 14$Ks_{A\&I}$. Analysing data regarding these three scopes of concerns in cost–benefit efficiency shows that the Pearson's correlation coefficients (0.561 between cost on registration and cost on implementation, 0.701 between cost and benefit of *ISO* 14$Ks_{A\&I}$) are significant at the 0.01 level (2-tailed). Frequencies of the three scopes of concerns on cost–benefit efficiency above mean grades are 50.0, 57.1, and 54.8%, all of which are above 50%. Moreover, a trend analysis between the cost–benefit efficiency and the ISO 14000 series' acceptability indicated that contractors who give higher score to the cost–benefit efficiency would have less intention to accept the ISO 14000 series. The survey results indicate that the indistinct cost–benefit efficiency obstructs the progression of the ISO 14000 series and contractors prefer to see a higher cost–benefit efficiency on the *ISO* 14$Ks_{A\&I}$.

Our survey results encounter another similar conclusion with the three previous surveys as detailed before in that contractors would hesitate to adopt and implement the ISO 14000 series if the cost is high. One way for small and medium-sized enterprises to reduce the cost is to form a network and establish a joint EMS in accordance with the ISO 14000 series. This route to achieve the ISO 14000 series has been proved effective at the Hackefors Industrial District in Sweden (Ammenberg *et al.* 2000).

## 2.5   The E+

### 2.5.1   Introduction

The EIA of construction projects is a process of identifying, predicting, evaluating, and mitigating the biophysical, social, and other relevant environmental effects of development proposals or projects prior to major decisions being taken and commitments made (IAIA 1997). According to the *Official Report on the State of the Environment in China 2001* (China EPB 2002), the annual implementation rate of EIA for construction projects was 97% in 2001 in mainland China. In addition, a further investigation on the implementation rate of EIA in mainland China indicates that the average EIA implementation rate from 1995 to 2001 is 88%, with an increasing rate of 23% (China EPB 2002).

On the other hand, three statistical figures from mainland China indicate that the EMS may not have been prevalent in the construction industry there; and they are given below.

- The first figure is the percentage of environmental certificates awarded to Chinese enterprises versus total environmental certificates awarded to enterprises worldwide, which is as low as 2% (ISO 2002);
- The second figure is the percentage of environmental certificates awarded to Chinese construction enterprises versus total environmental certificates awarded to Chinese enterprises, which is as low as 8% (ISO 2002);
- The third figure is the percentage of the construction enterprises that have been awarded environmental certificates versus total governmental registered construction enterprises in mainland China, which is as low as 0.1% (CACEB 2002; CEIN 2002).

It is obvious that implementation rate of EIA is much higher than the implementation rates of the ISO 14000 series in the construction industry in mainland China. These statistical figures also indicate that most construction enterprises have not yet adopted or accepted the ISO 14000 series in mainland China. Because of the disagreement between the implementation rates of EIA and EMS, there may be little coordination between the EIA process and EMS implementation in construction projects, and thus EIA may not really serve as a tool to promote EM in the construction industry in China. As a result, adverse environmental impacts such as noise, dust, waste, and hazardous emissions still occur frequently in construction projects in spite of their EIA approvals prior to construction.

However, this situation is expected to improve in the near future. The China Environmental Protection Bureau has enacted laws, in December 2002, to implement the environmental supervision system (ESS) in construction project management (China Environment Daily 16/12/2002). Although this supervision system had been carried out in 13 pilot construction projects only since 2002, it is suggested that contractors in mainland China have to pay greater attention to EM in project construction in future, and prepare to actually adopt and implement EM in construction in the near future.

To find out the main obstacles to implementing the ISO 14000 in the construction industry in mainland China, a questionnaire survey was conducted in 2001 among representative contractors in Shanghai, a representative city, and five key factors were identified. These five factors are (1) governmental command-and-control ordinances and regulations on encouraging contractors to adopt and implement EMS, (2) technology conditions for environment-friendly or resource-efficient construction, (3) competitive pressures from the competitors of both the domestic and international markets on adopting and implementing the ISO 14000 series, (4) cooperative attitude towards adopting and implementing the

ISO 14000, and (5) cost–benefit efficiency on adopting and implementing the ISO 14000 (Chen and Li *et al.* 2004b). According to the survey results, contractors in mainland China are most interested in technology conditions such as construction techniques and construction management approaches that can assist field engineers to reduce adverse environmental impacts in terms of the requirements of environmental ordinances and laws.

As can be seen from statistic figures and the questionnaire survey, the implementation of either the EMS or the ESS requires additional EM approaches as practicable as the EIA approach, which is popular and easier to use by contractors. For that reason, this chapter attempts to transplant a standard EMS process into a static EIA process, which is currently adopted in mainland China, to derive a dynamic EIA process. The EMS-based dynamic EIA process presented in this chapter, named as E+, is an integrative methodology which integrates practicable EM approaches into an ISO 14001 EMS process throughout a whole construction cycle in a construction project, and it is expected to be able to assist contractors to effectively and efficiently enhance their EM performance in China.

### 2.5.2 A conception model of the E+

The E+ is an integrative methodology for EM in construction projects, using which a dynamic EIA process can be effectively and efficiently applied during construction. The successful implementation of an EMS in construction projects requires far more than just the apparent prevention and reduction of adverse or negative environmental impacts in a new project and its construction process development cycles during pre-construction stage, continuous improvement of the EM function based on institutionalization of change throughout an onsite organization to reduce pollution during mid-construction stage, or efficient synergisms of pollution prevention and reduction such as waste recycle and regeneration in construction industry during mid-construction and post-construction stages. It necessitates a complete transformation of the construction management in an environmentally conscious enterprise, such as changes in management philosophy and leadership style, creation of an adaptive organizational structure, adoption of a more progressive organizational culture, revitalization of the relationship between the organization and its customers, and rejuvenation of other organizational functions (i.e. human resources engineering, research and development, finance, and marketing, etc.) (Azani 1999). In addition to the transformation for EM in construction enterprises, the integrative methodology, E+, for the effective implementation of EM in all phases of construction cycle including the pre-construction stage, the mid-construction stage, and the post-construction stage is necessarily activated, together with other rejuvenated construction management functions such as human resources, expert knowledge, and synergetic effect.

There are already some approaches to effectively implementing the EM onsite at different construction stages. For example, for the pre-construction stage, a CPI approach, which is a method to quantatively measure the amount of pollution and hazards generated by a construction process and construction project during construction, can be utilized by indicating the potential level of accumulated pollution and hazards generated from a construction site (Chen, Li and Wong 2000), and by reducing or mitigatingpollution level during the construction planning stage (Chen and Li *et al.* 2002); in addition to the CPI approach, a life-cycle assessment (LCA) approach for material selection (Lippiatt 1999), and a decision programming language (DPL) approach for environmental liability estimation (Jeljeli and Russell 1995), etc. also provide computable methods for making decision on EM at pre-construction stage; for the mid-construction stage, a crew-based incentive reward program (IRP) approach, which is realized by using bar-code system, can be utilized as an on-site material management system to control and reduce construction waste (Chen and Li *et al.* 2002a); for the post-construction stage, an online waste exchange (Webfill) approach which is further developed into an e-commerce system based on the trip-ticket system for waste disposal in Hong Kong can be utilized to reduce the final amount of C&D waste to be landfilled (disposed of the C&D waste in a landfill) (Chen and Li *et al.* 2003a). Although these approaches to EM in project construction have proved effective and efficient when they are used in a corresponding construction stage, it has also been noticed that these EM approaches can be further integrated for a total EM in construction based on the interrelationships among them. The integration can bring about not only a definite utilization of current EM approaches but also an improved environment for contractors to maximize the advantages of utilizing current EM approaches due to sharing EM-related information or data.

As mentioned above, the EMS is not as acceptable as EIA in mainland China partly due to the lack of efficient EM tools, and the tendency of EM in construction is to adopt and implement the EMS after the EIA report/form of a construction project is approved. As a result, the dynamic EIA process for contractors to enhance their environmental performance in mainland China, which integrates all necessary EM approaches available currently, just appropriates to the occasion.

The proposed E+ aims to provide high levels of insight and understanding regarding the EM issues related to the management in a construction cycle. In fact, current EIA process applied in mainland China is mainly conducted prior to the pre-construction stage of a construction project, when a contractor is required to submit an EIA report/form based on the size and significance of the project and the EIA process for the mid-construction stage is seldom conducted in normalized forms. Due to the alterability of the environmental impacts in the construction cycle, the commonly encountered static EIA process prior to construction cannot accommodate the implementation of the EMS in project construction, and a dynamic EIA process is thus designed for the E+. In addition, current EM approaches are to be combined with a frame of the EMS (a process of the EMS including issuing environmental policies, planning, implementation and

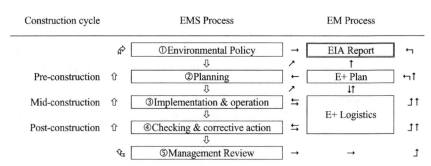

Legend: ⇧ = EM process flow, ↑ = EM data flow

Note: Description of the EMS Processes:
  ①*Environmental Policy*: the environmental policy and the requirements to pursue this policy via objectives, targets, and environmental programs;
  ②*Planning*: the analysis of the environmental aspects of the organization (including its processes, products and services as well as the goods and services used by the organization;
  ③*Implementation and operation*: implementation and organization of processes to control and improve operational activities that are critical from an environmental perspective (including both products and services of an organization);
  ④*Checking and corrective action*: checking and corrective action including the monitoring, measurement, and recording of the characteristics and activities that can have a significant impact on the environment;
  ⑤*Management Review*: review of the EMS by the organization's top management to ensure its continuing suitability, adequacy and effectiveness.

*Figure 2.2* A conception model of the E+.

operation, checking and corrective action, and management review) according to their interrelationships with which various EM-related information/data can be organized. Because the main task of the EM in construction is to reduce adverse environmental impacts, the dynamic data transference in the framework is the prime focus of the E+. Thus, a conception model of the E+ is illustrated in Figure 2.2.

## 2.6   Conclusions

The remarkable difference between the rate of ISO 14001 registration and EIA implementation indicates that contractors in mainland China have not really implemented EM and accepted the ISO 14000 series. This hypothesis has been tested in this study by a mail questionnaire survey conducted with contractors in Shanghai. The survey data has been analysed focusing on the ISO 14000 series' acceptability, and the survey results indicate that there are five classes (critical factors) affecting contractors in Shanghai on *ISO* $14Ks_{A\&I}$. These critical factors include governmental regulations, technology conditions, competitive pressures, cooperative attitude, and cost–benefit efficiency.

Based on the analysis of the ISO 14000 series' acceptability, an empirical evaluation model for deciding on whether to accept the ISO 14000 series has been developed (see Appendix B). The model can be used by contractors to decide whether they should accept the ISO 14000 series in the Shanghai construction industry.

The integrative methodology for EM in construction projects, in which a dynamic EIA process can be effectively and efficiently applied during construction, has been put forward. The implementation of the E+ model requires essential analytical approaches, which belong to the E+ Plan section or E+ Logistics section individually, to carry out data capture and transform stage by stage and realize its conclusive function.

# Chapter 3

# Effective prevention at pre-construction stage

## 3.1 Introduction

Environmental issues in construction typically include soil and ground contamination, water pollution, C&D waste, noise and vibration, dust, hazardous emissions and odours, demolition of wildlife and natural features, and archaeological destruction (Coventry and Woolveridge 1999). Since the early 1970s, there have been numerous studies related to environmental issues in construction. Some examples include the study on air pollution (Henderson 1970), noise pollution (U.S.EPA 1971, 1973), water pollution (McCullough and Nicklen 1971), and solid-waste pollution (Spivey 1974a,b) generated from construction sites. On the other hand, although the expression 'EM in construction' was first coined in the *U.S. National Environmental Policy Act* of 1969 (Warren 1973), the embryonic concept of EM in construction was not formulated until the late 1970s, when the role of environmental inspector was introduced in the design and construction phases of projects. The environmental inspector, who plays the role of environmental monitor (Dodds and Sternberger 1992), is a specialist whose academic background or experience results in considerable understanding of environmental impacts and applicable control measures, and acts as an advisor to construction engineers on all matters of EM (Spivey 1974a,b; Henningson 1978). Moreover, enthusiasm for establishing an EMS in a construction company increased quickly following two main important EM standards, BS 7750 (enacted in 1992) and the ISO 14001 EMS (enacted in 1996). The EM standards are regarded as guidance to the construction industry, from passive and one-sided CM on contamination reduction to active and all-round EM.

Pollution and hazards caused by construction projects have become a serious social problem all over the world. The sources of pollution and hazards from construction sites include dust, harmful gases, noises, blazing lights, solid and liquid wastes, ground movements, messy sites, fallen items, etc. These kinds of pollution and hazards can not only annoy residents nearby, but also affect the health and well-being of people in the entire city and area. For example, in big cities in developing countries, such as Shanghai and Beijing in mainland China, air quality has been deteriorating due to extensive and rapid urban redevelopment activities since the 1980s.

To tackle the serious environmental problems partly caused by construction pollution and hazards, environmental laws and regulations are increasingly enacted in different forms in different countries. For example, the Chinese government has issued a number of laws and regulations on environmental protection since the early 1980s. These laws and regulations include *Oceanic Environment Act* (enacted in 1982), *Water Pollution Protection Act* (enacted in 1984), *Air Pollution Protection Act* (enacted in 1987), and *Noise Pollution Protection Act* (enacted in 1989). Especially for the construction industry, the Chinese Ministry of Construction enacted the first *Construction Law* in 1998, which explicitly includes the liabilities and responsibilities of contractors in preventing and reducing the emission of pollutants to the natural environment; and the State Council of China enacted the *Managerial Ordinance on Environmental Protection of Construction Project* in the same year (SC of China 1998), which stipulates that all new construction projects must apply for environmental impact approval following an approval procedure of EIA report/form or EI form before construction. However, investigations by the authors of this book on many conflicts over construction pollution and hazards between construction practice and governmental regulations reveal that contractors need more effective, efficient, and economical EM tools to help them to obey all environmental laws and regulations.

As there are potential requirements of effective, efficient, and economical EM tools in the construction industry, this chapter aims to provide a systematic approach to dealing with environmental pollution potential generated in construction projects at pre-construction stage. The systematic approach comprises the CPI method to evaluate and reduce pollution and hazard levels of construction processes and construction projects, and the env.Plan method to quantitatively evaluate environmental-consciousness of proposed construction plans and thereby select the prime environment-friendly construction plan. This systematic approach allows for both qualitative analysis and control and quantitative assessments through measuring the CPI, and thus the selection of the prime environmental-conscious construction plan through env.Plan decision-making model. The authors believe that the qualitative assessment and control method is useful because it can provide construction project managers with essential knowledge of how to limit environmental pollution to its minimum at pre-construction stage. However, the systematic approach presented here is a necessary complement to EM in construction, as it can be adopted to quantitatively measure the degree of pollution and hazards generated in any particular construction processes and construction projects, then to re-arrange and revise construction plans and schedules in order to reduce the level of pollution and hazards, and thereafter to support decision-making on environmental-conscious construction.

## 3.2  CPI method

### 3.2.1  Qualitative analysis of construction pollution

The sources of pollution and hazards generated from construction activities can be divided into seven major types: dust, harmful gases, noise, solid and liquid wastes, fallen objects, ground movements, and others. In order to reduce and prevent these, it is necessary to identify first the construction operations that generate pollution and hazards. In Table 3.1, construction activities that generate pollution and hazards, and corresponding methods for prevention are listed. The contents in Table 3.1 are presented based on an extensive investigation on many construction cases, as well as numerous discussions with many project managers.

Qualitative methods to prevent pollution and hazards are divided into the following four categories:

1   Technology: This category recommends a range of advanced construction technologies which can reduce the amount of dust, harmful gases, noise, solid and liquid wastes, fallen objects, ground movements, and others. For example, replacing the impact hammer pile driver with the hydraulic piling machine can significantly reduce the level of noise generated by the piling operation.
2   Management: This category recommends the use of modern CM methods which may help reduce the amount of dust, noise, solid and liquid wastes, fallen objects, and others.
3   Planning: This category emphasizes revising and re-arranging construction schedules to reduce the aggregation of pollution and hazards. This category has effect on dust, noise, solid and liquid wastes, fallen objects, ground movements, and others.
4   Building material: Better building material can also help reduce pollution and hazards. This category has effect on harmful gases, fallen objects, ground movements, and others.

These four categories of preventive methods and their effects are also summarized in Table 3.2 (Chen, Li and Wong 2000).

The authors believe that it is possible to effectively control and reduce the amount of pollution and hazards in some respects by adopting these preventive methods. However, one limitation of the qualitative methods is their incapability towards quantifying and adjusting pollution and hazards of a construction procedure initiatively. In order to further quantitatively analyse the level of pollution and hazards, the next section describes a method to quantify and re-distribute pollution and hazards, generated from construction processes and construction projects, below legal limits.

*Table 3.1* Causes of pollution and hazards and preventive methods

| Type | Causes | Methods to prevent |
| --- | --- | --- |
| Dust | Demolition, rock blast | Static crushing/chemical breaking/water jet |
| | Excavation, rock drilling | Static crushing/chemical breaking/wet excavation/wet drilling |
| | Open-air rock power and soil | Covering/wet construction |
| | Open-air site and structure | Wet keeping/site clearing/mask |
| | Bulk material transportation | Awning/concrete goods/washing transporting equipment |
| | Bulk material loading and unloading | Concrete goods/packing and awning/wet keeping |
| | Open-air material | Awning/storehouse |
| | Transportation equipment | Cleaning |
| | Concrete and mortar making | Concrete goods |
| Harmful gases | Construction machine – pile driver | Hydraulic piling equipment |
| | Construction machine – crane | Electric machine |
| | Construction machine – electric welder | Bolt connection/pressure connection |
| | Construction machine – transport equipment | Night shift |
| | Construction machine – scraper | Electric machine |
| | Organic solvent | Poison-free solvent |
| | Electric welding | Bolt connection/pressure connection |
| | Cutting | Laser cutting |
| Noise | Demolition | Static crushing/chemical breaking |
| | Construction machine – pile driver | Hydraulic pile equipment |
| | Construction machine – Crane | Electric machine |
| | Construction machine – rock drill | Static crushing/chemical breaking |
| | Construction machine – mixing machinery | Concrete goods/prefabricated component |
| | Construction machine – cutting machine | Laser cutting machine/prefabrication/soundproof room/wall |
| | Construction machine – transport equipment | Night shift (based on the location of construction site) |
| | Construction machine – scraper | Night shift (based on the location of construction site) |

Table 3.1 (Continued)

| Type | Causes | Methods to prevent |
|------|--------|--------------------|
| Ground movements | Demolition | Static crushing/chemical breaking |
| | Pile driving | Static pressing-in pile |
| | Forced ramming | Static compacting/limited using |
| Wastes | Solid-state waste – building material waste | Prefabricated component/recovery |
| | Solid-state waste – building material package | Recovery |
| | Liquid waste – mud/building material waste | Recovery |
| | Liquid waste – machinery oil | Material saving |
| Fallen objects | Solid-state waste – building material waste | Material optimum seeking/technology improving |
| | Solid-state waste – building material package | Recovery |
| | Liquid waste – mud/Building material waste | Technology improving/recovery |
| | Liquid waste – construction water | Recovery |
| | Construction tools – scaffold and board | Safety control/reliable tools |
| | Construction tools – model plate | Technology improving/safety control |
| | Construction tools – building material | Technology improving/recovery |
| | Construction tools – sling/others | Safety control |
| Others | Urban transportation – road encroachment | Enclosing wall/night shift/underground construction |
| | Civic safety – demolition | Static crushing/chemical breaking |
| | Civic safety – automobile transportation | Overloading forbidden/speed limiting |
| | Civic safety – tower crane | Safety control |
| | Civic safety – construction elevator | Safety control |
| | Civic safety – foundation/earth dam | Safety control |
| | Urban landscape – structure exposed | Masking |
| | Urban landscape – night lighting | Using projection lamp |
| | Urban landscape – electric-arc light | Bolt connection/pressure connection/prefabricated component |
| | Urban landscape – mud/waste water | Drainage organization |
| | Urban landscape – civic facility destruction | Technology improving/plan preconception |

Table 3.2 Countermeasures for construction pollution and their effects

| Category | Pollution and hazards | | | | | | |
|----------|------|------------------|-------|-------------------|--------|-------------------|--------|
| | Dust | Harmful gases | Noise | Ground movements | Wastes | Fallen objects | Others |
| Technology | ✓ | ✓ | ✓ | ✓ | ✓ | ✓ | ✓ |
| Management | ✓ | x | ✓ | x | ✓ | ✓ | ✓ |
| Planning | ✓ | x | ✓ | x | o | x | ✓ |
| Material | x | ✓ | x | o | x | x | o |

Notes
✓ – More effective; o – Partial effective; x – Ineffective.

## 3.2.2   Construction pollution measurement

### 3.2.2.1   Pollution control in construction projects

Pollution control in construction projects can be defined as the control of all human activities that have either a significant or small negative impact on both natural and social environments during the entire construction process. It is an essential part of the implementation of EM in any individual construction project (Griffith *et al.* 2000). Construction pollution has been given great attention in the industry since the 1970s, not only in academic research but also in professional practice. From ASCE (www.asce.org), ICE (www.ice.org.uk), and EI (www.ei.org) online databases, the authors found that noise pollution inconstruction was first identified in a professional research in the early 1970s (U.S.EPA 1971), followed by air pollution (Jones 1973) solid-waste pollution (Skoyles and Hussey 1974; Spivey 1974a,b), and so forth. The concept of EM during construction was put forward in the late 1970s, and the role of environmental inspector, represented by a CM engineer, was introduced in the design and construction phases of projects. From then on, researches, worldwide, focused on the quantitative measurement and effective control approaches to reducing pollution and hazards, such as life-cycle costing; efficient energy consumption; reduction, re-use, and recycle of C&D material/debris; degradation and abatement of construction noise and dust; EIA, etc. Even so, there was little enthusiasm to establish an EMS in a commercial construction company until two main important standards, BS 7750 (1992) and the ISO 14001 EMS (1996), were promulgated. As the EMS is an organization's formal structure that implements EM (Griffith *et al.* 2000), approaches to construction pollution control are useful and effective in all environment-friendly practices in construction projects.

### 3.2.2.2   Construction pollution index

In many cases, conflicts between construction practice and governmental regulations arose regarding the permissible level of polluting emission, especially if the construction sites are in densely polluted areas. For example, the *Noise Pollution Protection Act* (NPPA 1993) in China specifies that the level of noise should not exceed 75 dB(A), above which site operations will be suspended by legal actions. In a construction site, the level of pollution emission from individual operations may not exceed the legal limits specified under the regulations; however, the aggregated level of pollution from multiple sources may exceed the limit. To prevent this and to ensure that the level of polluting emission does not exceed the legal limits during construction, a two-step quantitative method, as described in this section, can be followed. First, the method can predict the distribution of polluting-emission levels throughout a project's duration. Second, if it detects that the level of pollution exceeds the limit at a certain point of time, then on-site activities are re-scheduled so that the level of pollution can be re-distributed.

As a construction project generally spans over a year or even longer, the method of quantitative analysis should involve continuous monitoring and assessment for the entire project duration. CPI in measured as shown in Equation 3.1.

$$CPI = \sum_{i=1}^{n} CPI_i = \sum_{i=1}^{n} h_i \times D_i \tag{3.1}$$

where CPI is the construction pollution index of an urban construction project, $CPI_i$ is the CPI of a specific construction operation $i$, $h_i$ is the pollution and hazard magnitude per unit of time generated by a specific construction operation $i$, $D_i$ is the duration of the construction operation $i$ that generates pollution and hazards $h_i$, and $n$ is the number of construction operations that generate pollution and hazards.

In Equation 3.1, parameter $h_i$ is a relative variable, and its value is in the range of $[0, 1]$. If $h_i = 1$, it means that the pollution and hazards can cause fatal damage or catastrophes to people and properties nearby. For example, if a construction operation generates some noise and the sound level at the receiving end exceeds the "threshold of pain", which is 140 dB(A) (McMullan 1998), then the value of $h_i$ for this specific construction operation is 1. If $h_i = 0$, then it indicates that no pollution and hazards are detectable from a construction operation.

The initial value of each $h_i$ depends on experience and expert opinions and can be taken as the average of scores from experts. However, this calculation method cannot give an accurate value to each $h_i$ because the average may not be a real value of the $h_i$ or provide a most appropriate value to each $h_i$. To overcome this drawback in Equation 3.1, and to extend this quantitative pollution measurement approach from construction pollution indication to general P3 in construction and demolition projects, the authors developed an alternative index, i.e. stochastic process pollution index (SPPI) based on Equation 3.1. And it can be measured by Equations 3.2 and 3.3.

$$SPPI = \sum_{i=1}^{n} SPPI_i = \sum_{i=1}^{n} h_i \times D_i \tag{3.2}$$

$$h_i = \frac{h_i^{(optimistic)} + 4 \times h_i^{(mostlikely)} + h_i^{(pessimistic)}}{6} \tag{3.3}$$

where *SPPI* is the stochastic process pollution index of a project, $SPPI_i$ is the *SPPI* of a specific process $i$, $h_i$ is the expected hazard magnitude per unit of time generated by a specific process $i$, $h_i^{(optimistic)}$ is the optimistic hazard magnitude per unit of time generated by a specific process $i$, $h_i^{(mostlikely)}$ is the most likely hazard magnitude per unit of time generated by a specific process $i$, $h_i^{(pessimistic)}$ is the pessimistic hazard magnitude per unit of time generated by a specific process $i$, $D_i$ is the duration of the specific process $i$ that generates pollution and/or hazard $h_i$, $n$ is the number of processes that generate pollution and hazards.

Equations 3.2 and 3.3 provide an innovative way to define $h_i$. The SPPI assumes a beta probability distribution for the $h_i$ estimates. Regarding each $h_i$, each expert will provide a set of values – $h_i^{(optimistic)}$, $h_i^{(mostlikely)}$, and $h_i^{(pessimistic)}$ – from which the expected $h_i$ is calculated by their weighted average. Comparing with the programme evaluation and review technique (PERT) adopted in project scheduling, the approximate treatment gives a more reasonable result for each $h_i$.

It is then possible to identify values of $h_i$ for all types of pollution and hazards generated by commonly used construction operations. For example, according to the information on sound emission from piling machines, as well as the types of piles, the authors derive the values of $h_i$ for some piling operations (Table 3.3).

Data such as those regarding the emissions of noise, harmful gases, and wastes are normally available in the product specifications of construction machinery and equipment, or can be conveniently measured. These data can then be converted to $h_i$ value by normalizing them into the range of [0, 1]. In case there is no data available for such conversion, $h_i$ values have to be decided based on the user's experience and expert opinions.

It is also very useful to create a CPI bar chart. The CPI bar chart is very similar to the ordinary bar charts used in construction scheduling, except that the thickness of the bars in the histogram represents the $h_i$ value for the corresponding construction operation. By integrating the concept of CPI method into Microsoft Project©, which is a commonly used tool in construction project management, the authors think it is possible to develop a system to neatly combine EM with project management, as shown in Figure 3.1.

*Table 3.3* Values of $h_i$ for some piling operations

| Number | Piling operations | $h_i$ value (per day) |
|---|---|---|
| 1 | Prefabricated concrete piles using drop-hammer driver | 0.5 |
| 2 | Sheet steel piles using drop-hammer driver | 0.6 |
| 3 | Prefabricated concrete piles using hydraulic piling driver | 0.2 |
| 4 | Sheet steel piles using hydraulic piling driver | 0.3 |
| 5 | Bored piling | 0.1 |
| 6 | Sheet steel piles using drop-hammer driver | 0.7 |
| 7 | Prefabricated concrete piles using static pressing-in driver | 0.2 |

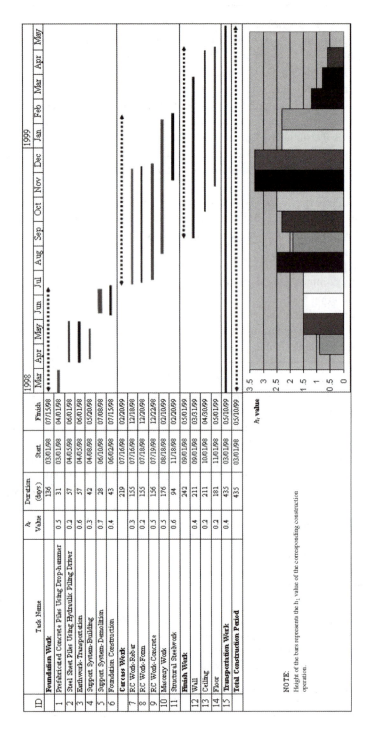

*Figure 3.1* Project scheduling together with EM using CPI.

In Figure 3.1, the $h_i$ values are listed next to the task name of their corresponding construction operations. As the height of the bar represents the $h_i$ value, the area of the bar represents the CPI value of the corresponding construction operation. For example, the sample construction project (refer to Figure 3.1) involves a piling operation which includes the following activities and durations (measured in number of days):

- Driving prefabricated concrete piles using drop-hammer driver, and duration is 31 days.
- Driving sheet steel piles using hydraulic piling driver, and duration is 57 days.

Then, according to Equation 3.1, the value of CPI for the piling operation is $0.5 \times 31 + 0.3 \times 57 = 32.6$, and the overall CPI value for the project is 747.2. The value of CPI reflects the accumulated amount of pollution and hazards generated by a construction project within its project duration. That is the aggregation of the thickness of histograms, as indicated at the bottom of the bar chart (see Figure 3.1), represents the distribution of the CPI value along the whole project duration. This distribution is particularly useful for project managers to identify the periods when the project will generate the highest amount of pollution and hazards. Therefore, preventive methods such as those listed in Table 3.1 can be applied to reduce the amount of pollution and hazards during those periods.

Careful study of the sample project revealed that during November–December 1998 the project generated the highest pollution and hazard level according to the distribution diagram of Figure 3.1, and the root of the pollution is the large amount of on-site mixing of concrete and masonry work during that period. The project manager foresaw the problem, and decided to reduce the amount of on-site mixing of concrete in those months by using 25% ready-mixed concrete. As a result, the amount of noise generated from on-site concrete mixing was reduced. The $h_i$ value decreased in November–December 1998 from 3.3 to 2.5, a 25% reduction in the value of $h_i$. It also indicates that the total amount of pollution and hazards is consequently reduced.

Figure 3.2 illustrates another example of a construction project comprising 20 activities. The $h_i$ value of each activity is presented in Table 3.4 and indicated at the right side of the bars in Figure 3.2. For example, the $h_i$ value for "RC Formwork" is calculated to be 0.5. Moreover, the y-axis in Figure 3.3 represents the accumulated $h_i$ value and the x-axis is for the project duration. Thus, the shaded area is the total CPI value. It is suggested that the maximum permissible level of $h_i$ is 0.8 at any point of time during construction. It is necessary to note that the definition of maximum level of $h_i$ value is based on the average allowable pollution and hazard level. The value of maximum $h_i$ can be adjusted to reflect the level of pollution and hazard control: the lower the maximum $h_i$ value, the tighter the control on pollution and hazards, and vice versa.

| Task Name | Duration | Priority | Predecessors | Aug Sep Oct Nov Dec Jan Feb Mar Apr May Jun Jul A |
|---|---|---|---|---|
| 1 | Demolition | 6 d | 0 | | hi=0.8 |
| 2 | Site Preparation | 6 d | 0 | 1 | hi=0.8 |
| 3 | Cast-In-Place RC Pile | 20 d | 0 | 2 | hi=0.5 |
| 4 | Excavation & Support System | 30 d | 0 | 3 | hi=0.8 |
| 5 | Foundation Baseplate | 6 d | 0 | 4 | hi=0.5 |
| 6 | RC Formwork | 42 d | 0 | 5 | hi=0.5 |
| 7 | Steel Formwork | 30 d | 0 | 6 | hi=0.2 |
| 8 | Roof works | 6 d | 0 | 7 | hi=0.5 |
| 9 | Water supply & sewerage works | 30 d | 0 | 7 | hi=0.1 |
| 10 | Power supply system | 30 d | 0 | 7 | hi=0.1 |
| 11 | Lighting system | 20 d | 0 | 7 | hi=0.1 |
| 12 | Air Conditioning | 30 d | 0 | 7 | hi=0.1 |
| 13 | Computer & communication network | 30 d | 0 | 7 | hi=0.1 |
| 14 | Floor finish & polishing | 50 d | 0 | 8 | hi=0.7 |
| 15 | Internal wall finish | 30 d | 0 | 14 | hi=0.4 |
| 16 | External wall finish | 20 d | 0 | 8 | hi=0.2 |
| 17 | Internal partition wall | 30 d | 0 | 9,10,11,12,13 | hi=0.1 |
| 18 | Ceiling work | 40 d | 1000 | 15 | hi=0.2 |
| 19 | Site improvements | 6 d | 0 | 18 | hi=0.2 |
| 20 | Landscaping work | 6 d | 0 | 19 | hi=0.1 |

*Figure 3.2* Initial schedule of a construction project.

*Table 3.4* $h_i$ values of some construction operations

| Task name | $h_i$ Value (per day) |
|---|---|
| Demolition | 0.7 |
| Site preparation | 0.7 |
| Cast-in-place RC Pile | 0.5 |
| Excavation and support system | 0.7 |
| Foundation baseplate | 0.3 |
| RC framework | 0.5 |
| Steel framework | 0.2 |
| Roof works | 0.5 |
| Water supply and sewerage works | 0.1 |
| Power supply system | 0.1 |
| Lighting system | 0.1 |
| Air conditioning | 0.1 |
| Computer and communication network | 0.1 |
| Floor finish and polishing | 0.7 |
| Internal wall finish | 0.4 |
| External wall finish | 0.2 |
| Internal partition wall | 0.1 |
| Ceiling work | 0.2 |
| Site improvements | 0.2 |
| Landscaping work | 0.1 |

It is also necessary to note the CPI histogram is produced by linearly accumulating $h_i$ values. This may cause inaccuracies as some pollution measurements such as noise levels cannot be linearly added up. The authors are examining, at the time of writing, the effect of nonlinearity and are aiming to develop a revised method to accumulate $h_i$ values so that accurate histograms

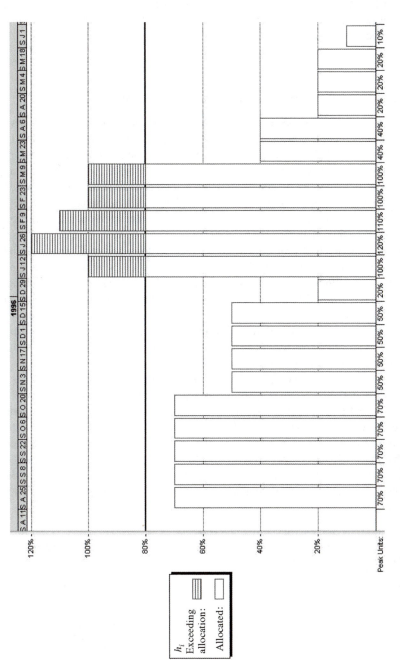

Figure 3.3 Histogram of $h_i$ in the initial schedule.

can be produced. However, it can be seen from Figure 3.2 that during the period December 1996 to March 1997 of the project duration, the level of $h_i$ values will exceed the maximum value, indicating that during this period, the accumulated level of pollution will exceed the limit. Therefore, it is necessary to re-arrange the project schedule so that the level of pollution can be reduced below the limit.

### 3.2.3 A pseudo-resource approach for CPI levelling

Resource levelling can smooth daily resource demands, and it is an effective tool for construction project scheduling when construction resource conflicts or shortages occur. This section presents a method to combine pollution and hazard control with traditional construction resource levelling at project scheduling stage. The $h_i$ values are treated as a pseudo resource, and the maximum $h_i$ value is treated as the limit of the pseudo resource. This pseudo resource together with other types of resources can be levelled by using the traditional construction resource levelling methods (Pilcher 1992).

In the experimental project schedule, which is described in Figures 3.2 and 3.3, the authors found that if they set $h_i$ as a kind of pseudo resource, then construction pollution and hazards can be levelled following resource levelling. Although there would be different construction pollution emissions depending on the different daily resources demanded in a schedule, it is still possible to adjust the level of construction pollution with the help of $h_i$. As $h_i$ is a measurement relative to all other real resources such as materials and workers in a schedule, it can be integrated with resource optimization.

In the sample project considered, there are six kinds of construction resources – workers, materials, machines, instruments, and power (denoted as $R_1, R_2, R_3, R_4$, and $R_5$); and pollution and hazards from construction are treated as a pseudo resource, which is denoted as $R_6$. These resources are listed in Table 3.5. For the purpose of convenience in calculation, the values of the resources are adjusted so that there will be no very large or small figures.

In order to test the pseudo-resource approach, the authors chose Microsoft Project© as a tool for scheduling and resource levelling. The project schedules

Table 3.5 Resources in initial construction schedule

| Resource name | Mark | Max units available | Adjustment |
|---|---|---|---|
| Workers | $R_1$ | 1900 | Workers no. × 10 |
| Materials | $R_2$ | 2200 | Materials cost × 0.01 |
| Machines | $R_3$ | 2100 | Machines cost × 0.01 |
| Instruments | $R_4$ | 3100 | Instruments cost × 0.01 |
| Power | $R_5$ | 3400 | Power cost × 0.01 |
| $h_i$ | $R_6$ | 80 | CPI × 100 |

| | Task Name | Duration | Priority | Predecessors |
|---|---|---|---|---|
| 1 | Demolition | 6 d | 100 | |
| 2 | Site Preparation | 6 d | 100 | 1 |
| 3 | Cast-In-Place RC Pile | 20 d | 100 | 2 |
| 4 | Excavation & Support System | 30 d | 100 | 3 |
| 5 | Foundation Baseplate | 6 d | 100 | 4 |
| 6 | RC Formwork | 42 d | 100 | 5 |
| 7 | Steel Formwork | 30 d | 100 | 6 |
| 8 | Roof works | 6 d | 100 | 7 |
| 9 | Water supply & sewerage works | 30 d | 100 | 7 |
| 10 | Power supply system | 30 d | 100 | 7 |
| 11 | Lighting system | 20 d | 100 | 7 |
| 12 | Air Conditioning | 30 d | 100 | 7 |
| 13 | Computer & communication network | 30 d | 100 | 7 |
| 14 | Floor finish & polishing | 50 d | 100 | 8 |
| 15 | Internal wall finish | 30 d | 100 | 14 |
| 16 | External wall finish | 20 d | 100 | 8 |
| 17 | Internal partition wall | 30 d | 100 | 9,10,11,12,13 |
| 18 | Ceiling work | 40 d | 900 | 15 |
| 19 | Site improvements | 6 d | 100 | 18 |
| 20 | Landscaping work | 6 d | 100 | 19 |

*Figure 3.4* Microsoft Project©-levelled project schedule.

levelled by Microsoft Project© and the corresponding histogram of $h_i$ values are illustrated in Figures 3.4 and 3.5, respectively. From Figures 3.4 and 3.5, we find that the construction pollution level spreads out under the line of the maximum permissible level of $h_i$ (maximum $h_i = 0.8$) when the other five resources (refer to Table 3.5) are also levelled down to their individual resource limits. Therefore, the pseudo-resource approach for reducing construction pollution and hazard level is feasible at project scheduling stage. However, the total construction period is stretched by 22 days in Figure 3.4 after resource levelling. It is about 8% longer than the original schedule in Figure 3.2. Similar results were obtained from additional experimental schedules, which are not presented in this book. The experimental research therefore revealed that the pseudo-resource approach can assist project managers to keep construction pollution and hazard level below a legal range while making little difference to their normal schedules. The results from the experiment also indicated that it is necessary to find an optimum method to arrive at a shorter schedule for the proposed construction project with every resource levelled, including the pseudo resource.

### 3.2.4   CPI levelling using GA

A comparative analysis of resource-levelling and resource-allocation capabilities of project management software packages indicates that heuristic methods have a better performance than Microsoft Project© and Primavera Project Planner (Farid and Manoharan 1996). In recent years, research on construction schedule has improved the theory of resource levelling and allocation with heuristic techniques (Reeves 1993) considerably. For example, an artificial neural network (ANN) is used to minimize project duration and cost by using a mathematical model based

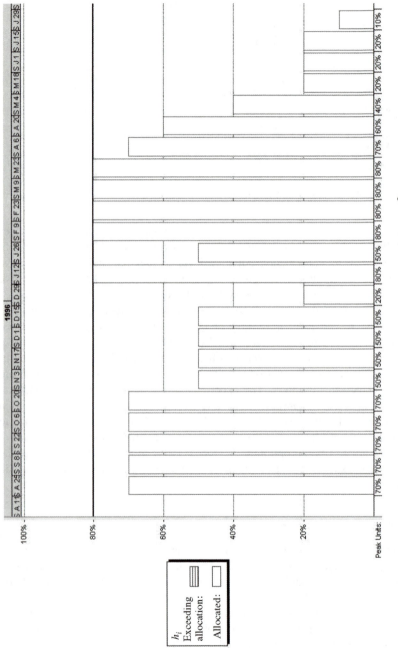

Figure 3.5 Histogram of $h_i$ values associated with the schedule levelled by Microsoft Project©.

on precedence relationships, multiple crew-strategies, and time–cost trade-off (Adeli and Karim 2001; Senouci and Adeli 2001), and GA is used to search for a near-optimum solution to the problem of resource allocation and levelling integrated with time–cost trade-off model, resource-limited/constrained model, and resource levelling model (Chan *et al.* 1996; Chua *et al.* 1996; Li and Love 1997; Li, Cao, and Love, 1999; Hegazy 1999; Leu and Yang 1999; Leu *et al.* 1999). To integrate various heuristic methods into resource levelling, the methods used by Harris (1978) and Hegazy (1999), which minimize both daily fluctuations in resource use and the resource utilization period, have been adapted. According to Hegazy (1999), the moment of fluctuations in daily resource use can be calculated using Equation 3.4.

$$M_x^R = \sum_{j=1}^{n} RP_j^2 \tag{3.4}$$

And the moment for measuring the resource utilization period is calculated using Equation 3.5.

$$M_y^R = \sum_{j=k}^{n} (j-k) RP_j \tag{3.5}$$

These two moment calculations can be used in minimizing either resource fluctuations or the duration of resource use, or both. As concurrent optimization of resource levelling and pollution and hazard control is a nonlinear searching problem, GA is suitable to solve it.

### 3.2.4.1 Gene formation

In a number of commercial resource levelling software packages, the user is allowed to set priority levels to tasks. Priority is an indication of a task's importance and availability for levelling (that is, resolving resource conflicts or over-allocations by delaying or splitting certain tasks). The task priority setting controls levelling, which allows users to control the order in which software systems such as Microsoft Project[©] can delay tasks with over-allocated resources. Tasks with the lowest priority are delayed or split first, and tasks with a higher priority are not levelled before other tasks sharing the over-allocated resources. A previous comparison of heuristic and optimum solutions in resource-constrained project scheduling shows activity priority to be a key factor of a heuristic rule. The heuristic rule which bases activity priority on activity slack produced an optimal schedule span most and exhibited the lowest average increase above optimum of the heuristic rules examined (Davis and Patterson 1975). A heuristic fuzzy expert system has also proved that priority ranking can obtain an optimum result in construction resource scheduling (Chang *et al.* 1990). Thus, to apply GA to solve the multiple-resources levelling problem, it is essential to have a

| 1 | 2 | 3 | 4 | 5 | 6 | 7 | 8 | 9 | | j |
|---|---|---|---|---|---|---|---|---|---|---|
| $P_1$ | $P_2$ | $P_3$ | $P_4$ | $P_5$ | $P_6$ | $P_7$ | $P_8$ | $P_9$ | ... | $P_j$ |

Notes
1. $P_j$ is the priority of active $j$, $0 \leq P_j \leq 8$.
   $P_j = 0$, activity priority is highest;
   $P_j = 1$, activity priority is higher;
   $P_j = 2$, activity priority is very high;
   $P_j = 3$, activity priority is high;
   $P_j = 4$, activity priority is medium;
   $P_j = 5$, activity priority is low;
   $P_j = 6$, activity priority is very low;
   $P_j = 7$, activity priority is lower;
   $P_j = 8$, activity priority is lowest.
2. The priority values are in accordance with the priority grades of actives in Microsoft Project®.

*Figure 3.6* Gene formation (Hegazy 1999).

gene structure that facilitates the operations of GA. Bearing this in mind, the following gene format used by Syswerda and Palmucci (1991), Grobler *et al.* (1995), Boggess and Abdul (1997), and Hegazy (1999) has been adopted:

In Figure 3.6, a string has $j$ genes, and each box represents a gene. The number inside the box is the priority setting for a particular task labelled by the number above the box. A string is a particular combination of priority settings that determines a specific schedule. The fitness of the string is evaluated by the following function (Hegazy 1999),

$$\omega_d(D_i/D_0) + \sum_{j=1}^{n} [\omega_j^R(M_{xji}^R + M_{yji}^R)/(M_{xj0}^R + M_{yj0}^R)] \tag{3.6}$$

where $M_x^R$ is the moment of fluctuations of daily resource use as defined in (3.4); $M_{xji}^R$ is the moment of fluctuations of resource use in a specific schedule determined by string $i$ in day $j$; $M_{xj0}^R$ is the initial value of $M_x^R$ in day $j$; $M_y^R$ is the moment of resource utilization period, as defined in (3.5); $M_{yji}^R$ is the moment of resource utilization period of a schedule determined by a string $i$ in day $j$; $M_{yj0}^R$ is the initial value of $M_y^R$ in day $j$; $D_i$ is the new project duration of the schedule determined by string $i$; $D_0$ is the initial project duration determined by any resource allocation heuristic rule; $\omega_d$ is the weight in minimizing project duration; $\omega_j^R$ is the weight in levelling every resource in day $j$; $i$ is the generation number of genes; $j$ is the representative day during a project's total working days, and $n$ is the number of working days in a project's duration.

By selecting different weights, the fitness function (3.6) enables the user to conduct different heuristics-based resource levelling including reducing resource fluctuations, minimizing the duration of resource use, or both.

### 3.2.4.2 Experimental results

This section presents experimental results obtained by using GA to combine pollution control and resource allocation into the task of resource levelling. The schedule used in the experiment is that of a construction project in Shanghai, in which there are 20 activities for general control, and the initial schedule of the activities and their associated level of polluting emission ($h_i$ value) are shown in Figure 3.2. From the histogram of $h_i$ values, which is illustrated in Figure 3.3, it can be seen that the accumulated level of polluting emission exceeds the permissible limit.

In the experiment, the initial population size is set at 100. Also, to minimize both resource fluctuations and duration, the weightings in the fitness function (3.6) are given an equal weighting of 1. The resultant schedule and the associated histogram of $h_i$ values are illustrated in Figures 3.7 and 3.8.

Comparing the GA-levelled schedule with the Microsoft Project©-levelled schedule, it can be seen that the priorities of resource use in the GA levelled schedule are set at different values (Figure 3.7); whereas priorities in the Microsoft Project©-levelled schedule (Figure 3.4) do not have any changes from the original schedule (Figure 3.2). In addition, the duration of the GA-levelled schedule is 298 days, which is shorter than the duration of the schedule levelled by Microsoft Project© (302 days). Moreover, two additional experiments conducted by the authors also support these facts. From the experiments, the authors conclude that the GA can adjust the task priorities for the re-distribution of resources to meet resources constraints and make the schedule shorter; moreover, the GA enhances the levelling function of Microsoft Project©, as it enables the user to identify the optimal settings of task priorities in resource levelling automatically.

| | Task Name | Duration | Priority | Predecessors |
|---|---|---|---|---|
| 1 | Demolition | 6 d | 100 | |
| 2 | Site Preparation | 6 d | 100 | 1 |
| 3 | Cast-In-Place RC Pile | 20 d | 100 | 2 |
| 4 | Excavation & Support System | 30 d | 100 | 3 |
| 5 | Foundation Baseplate | 6 d | 100 | 4 |
| 6 | RC Formwork | 42 d | 100 | 5 |
| 7 | Steel Formwork | 30 d | 100 | 6 |
| 8 | Roof works | 6 d | 100 | 7 |
| 9 | Water supply & sewerage works | 30 d | 100 | 7 |
| 10 | Power supply system | 30 d | 100 | 7 |
| 11 | Lighting system | 20 d | 100 | 7 |
| 12 | Air Conditioning | 30 d | 100 | 7 |
| 13 | Computer & communication network | 30 d | 100 | 7 |
| 14 | Floor finish & polishing | 50 d | 100 | 8 |
| 15 | Internal wall finish | 30 d | 100 | 14 |
| 16 | External wall finish | 20 d | 100 | 8 |
| 17 | Internal partition wall | 30 d | 100 | 9,10,11,12,13 |
| 18 | Ceiling work | 40 d | 900 | 15 |
| 19 | Site improvements | 6 d | 100 | 18 |
| 20 | Landscaping work | 6 d | 100 | 19 |

*Figure 3.7* GA-optimized construction schedule.

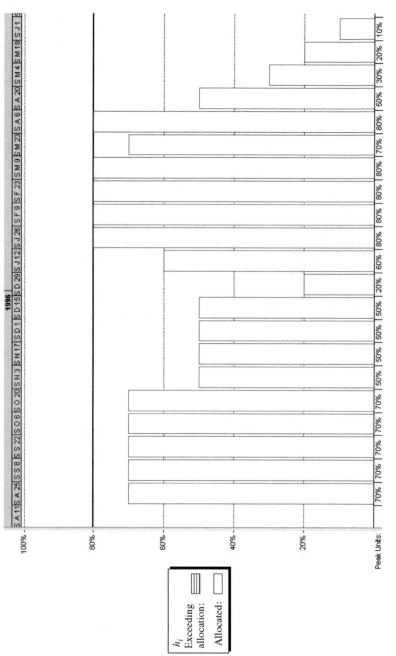

*Figure 3.8* Histogram of $h_i$ values associated with the schedule levelled by GA.

## 3.3    Env.Plan method

### 3.3.1    Introduction

Although the CPI method has demonstrated its effectiveness and usefulness in indicating, reducing, or mitigating pollution and hazard level during construction planning stage (Chen *et al.* 2000; Li *et al.* 2002), the problem of how to select the best construction plan based on levelling the magnitude of quantified adverse environmental impacts of construction operations is still a research task. Moreover, the major premise of CPI's application in construction plan evaluation is that each construction activity's CPI can be linearly aggregated, and this hypothesis cannot directly reflect the complicated nonlinear causal relationship among construction activities that have environmental impact. In this section, the authors introduce the use of ANP to develop a decision support model named env.Plan. This method aims to integrate important considerations of construction planning, which includes time, cost, quality, and safety, with the evaluation of the impact of various environmental factors, so that the most suitable plan can be obtained.

A construction plan is normally evaluated through fixed criteria such as cost, time, quality, safety, and so on during the planning period. Since effective planning has considerable influence on the successful completion of a construction project, both construction managers and researchers are aware of tools used to prepare and evaluate a construction plan. The analytic hierarchy process (AHP), which is known as a powerful and flexible decision-making process to help people set priorities and make the best decision when both qualitative and quantitative aspects of a decision need to be considered, has been utilized in various areas of construction research and practice since the late 1970s (Zeeger and Rizenbergs 1979), including construction planning (Dey *et al.* 1996). In this regard, the AHP method is recommended by construction researchers as a useful multicriteria assessment tool for its stronger mathematical foundation, its ability to gauge consistency of judgements, and its flexibility in the choice of ranges at the subcriteria level (Khasnabis *et al.* 2002).

However, a notable weakness of AHP is that it cannot deal with interconnections between decision factors in the same level, because an AHP model is structured in a hierarchy in which no horizontal links are allowed. In fact, this weakness can be overcome by using a senior multicriteria analytical technique known as ANP. The ANP is more powerful in modelling complex decision environments than the AHP because it can be used to model very sophisticated decisions involving a variety of interactions and dependencies (Meade and Sarkis 1999; Saaty 1999). These advantages are embodied in several examples of applications of the ANP (Srisoepardani 1996). For example, Saaty (1996) recommended the ANP to be used in cases where the most thorough and systematic analysis of influences needs to be made. In addition, the ANP method has been successfully applied to the strategic evaluations of environmental practices and programmes in both manufacturing and business to help analyse various project-, technological- or business-decision alternatives, and it also has been proved to be useful for modelling dynamic strategies and systemic influences on managerial decisions related to the EM (Meade and Sarkis 1999). As a result, the ANP is selected.

### 3.3.2  Environmental indicators

In order to find suitable environmental indicators to evaluate a construction plan, the authors conducted an extensive literature review according to a classification of environmental indicators. The literature review on environmental issues in construction was conducted in several dominant databases. These are the Civil Engineering Database (CEDB) of the American Society of Civil Engineers (ASCE), the Compendex® database of the Engineering Index (EI), the Engineering News-Record (ENR) executive search engine (enr.com) and magazines of the McGraw-Hill Companies, the Construction Plus (CN+) search engine (www.cnplus.co.uk) of the Emap Construction Network, and the advanced search engine of the U.S. Environmental Protection Agency (U.S.EPA) (www.epa.gov). In addition to these five dominant databases, a commonly used search engine, Google (www.google.com), was also employed to search for online literature. The search results contained thousands of articles and reports related to environmental impacts and EM in construction practice.

A summary of literature retrieved is listed in Table 3.6. This included 367 references in the ASCE's CEDB and 908 references in the EI's Compendex®, which are relevant to environment-friendly technology, management, and material.

Environmental indicators here refer to factors in a construction project that can adversely or favourably impact on the natural environment and can directly influence construction planning. Based on this, environmental factors can be grouped into adverse environmental factors (denoted as EA factors) and favourable environmental factors (denoted as EF factors). The third category of indicators is those that may lead to adverse or favourable environmental impact depending on the specific environmental conditions in which a construction project is executed. This category of environmental indicators is named as uncertain environmental indicators, or EU factors.

Following the classification described above, a procedure for identifying environmental indicators is illustrated in Figure 3.9. It indicates that the environmental indicators were identified based on an extensive literature review of databases and online materials. The environmental indicators are interrelated with technology, resource, time, cost, management, society, and the natural environment in which a construction project is executed.

Environmental indicators for construction planning are identified and sorted by their environmental impacts ($EI_i$) in Table 3.7. The value of environmental impacts for each environmental indicator $i(EI_i)$ is calculated using Equation 3.7, which is a sum of eight generally recognized but most serious environmental hazards caused by the indicator. These eight hazards include soil and ground

*Table 3.6* A statistical classification of referred articles on environmental issues

| Research highlight | Reference and starting point | Reference amount (as of 31/12/2002) | |
|---|---|---|---|
| | | ASCE's CEDB (since 1972) | EI's Compendex® (since 1970) |
| Technology | | 94 | 358 |
| Environment-friendly innovative technology | Taylor et al. 1976 | 36 | 65 |
| Pollution prevention and minimization | | 58 | 293 |
|   Air pollution | Henderson 1970 | – | – |
|   Noise pollution | U.S.EPA 1971 | – | – |
|   Water pollution | McCullough and Nicklen 1971 | – | – |
|   Waste pollution | Spivey 1974a,b | – | – |
| Management | | 213 | 367 |
| Environmental survey | Spivey 1974a,b | 12 | 41 |
| Environmental/Quality management system | Dohrenwend | 11 | 28 |
| Environmental/Quality management approach | Dohrenwend 1973 | 7 | 18 |
| Information technology | Kawal 1971 | 183 | 280 |
| Material | | 60 | 183 |
| Eco-friendly regenerated construction material | Emery 1974 | 35 | 93 |
| Waste re-use and recycling | Spivey 1974a,b | 25 | 90 |

Notes
1 ASCE's CEDB is available online via http://www.pubs.asce.org/cedbsrch.html;
2 EI's Compendex® is available online via http://www.engineeringvillage2.org/.

contamination, ground and underground water pollution, C&D waste, noise and vibration, dust, hazardous emissions and odours, impacts on wildlife and natural features, and archaeological impacts (Chen, Li and Wong 2000).

$$EI_i = \sum_{j=1}^{8} EI_{i,j} \quad (j = 1, 2, \ldots, 8)$$

(3.7)

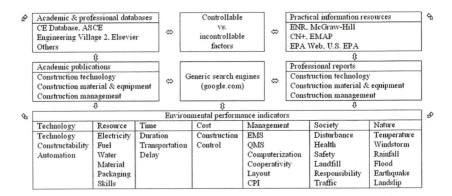

| Academic & professional databases | | | Controllable | | Practical information resources | | |
| CE Database, ASCE | | | vs. | | ENR, McGraw-Hill | | |
| Engineering Village 2, Elsevier | | | incontrollable | | CN+, EMAP | | |
| Others | | | factors | | EPA Web, U.S. EPA | | |

| Academic publications | | | | | Professional reports | | |
| Construction technology | | | Generic search engines | | Construction technology | | |
| Construction material & equipment | | | (google.com) | | Construction material & equipment | | |
| Construction management | | | | | Construction management | | |

Environmental performance indicators

| Technology | Resource | Time | Cost | Management | Society | Nature |
|---|---|---|---|---|---|---|
| Technology | Electricity | Duration | Construction | EMS | Disturbance | Temperature |
| Constructability | Fuel | Transportation | Control | QMS | Health | Windstorm |
| Automation | Water | Delay | | Computerization | Safety | Rainfall |
| | Material | | | Cooperativity | Landfill | Flood |
| | Packaging | | | Layout | Responsibility | Earthquake |
| | Skills | | | CPI | Traffic | Landslip |

*Figure 3.9* The framework for identifying environmental indicators.

where $EI_i$ is the total environmental impact caused by environmental indicator $i$, and $EI_{i,j}$ is individual environmental impact caused by eight possible hazards including soil and ground contamination ($j = 1$), ground and underground water pollution ($j = 2$), C&D waste ($j = 3$), noise and vibration ($j = 4$), dust ($j = 5$), hazardous emissions and odours ($j = 6$), impacts on wildlife and natural features ($j=7$), and archaeological impacts ($j = 8$) caused by the environmental indicator $i$. Its value is defined to be one of the three choices $\{-1, 0, +1\}$; where $-1$ represents that the environmental indicator will intensify the level of hazards, 0 represents that the effect of the environmental indicator is uncertain, and $+1$ represents that the indicator can reduce the level of hazards.

The assumed value of environmental impact of each environmental indicator ($EI_i$) is then used to reclassify the environmental indicators which have been identified from the literature review so that the new classification can be more flexible to all kinds of construction projects. The environmental indicators, with their original classification, and corresponding values of $EI_{i,j}$ are listed in Table 3.7. According to the results of environmental impacts listed in Table 3.7, all environmental indicators are finally classified into EA Factors ($EI_i < 0$), EF Factors ($EI_i > 0$), and EU Factors ($EI_i = 0$) (refer to Table 3.8). These reclassified environmental indicators are to be used for constructing an ANP model for evaluating environmental impact of a construction plan.

In addition to the classification of these environmental indicators and their $EI_i$ values, Table 3.8 also provides corresponding values of experimental plan alternatives Plan A, Plan B and Plan C, based on a construction background in Shanghai, China.

### 3.3.3 ANP model and approach

As defined by Saaty (1996/1999), the ANP is a general theory of relative measurement used to derive composite priority ratio scales from individual

Table 3.7 Environmental indicators and their potential environmental impacts as to a construction plan

| Class | Environmental indicators | Unit | Potential environmental impacts (EI_i) | | | | | | | | | Representative references |
|---|---|---|---|---|---|---|---|---|---|---|---|---|
| | | | $EI_{i,1}$ | $EI_{i,2}$ | $EI_{i,3}$ | $EI_{i,4}$ | $EI_{i,5}$ | $EI_{i,6}$ | $EI_{i,7}$ | $EI_{i,8}$ | $\sum_{j=1}^{8} EI_{i,j}$ | |
| Technology | Cleaner technologies and Automation ratio | % | +1 | +1 | +1 | +1 | +1 | +1 | +1 | +1 | +8 | Rosenfeld and Shapira 1998; Jones and Klassen 2001; Tiwari 2001; Reddy and Jagadish 2003 |
| | Constructability | % | 0 | 0 | 0 | 0 | 0 | 0 | 0 | 0 | 0 | Mifkovic and Peterson 1975; Hinckley 1986; Bonforte and Keeber 1993 |
| Resource | Electricity consumption amount | kWh | 0 | 0 | −1 | −1 | −1 | −1 | 0 | 0 | −4 | Hendrickson and Horvath 2000 |
| | Fuel consumption amount | joule | −1 | −1 | −1 | −1 | −1 | −1 | −1 | −1 | −8 | Mohr 1975; Peyton 1977; Reardon 1995; Peurifoy 2002 |
| | Water consumption amount | ton | −1 | −1 | −1 | 0 | +1 | 0 | −1 | 0 | −4 | Gambatese and James 2001 |
| | Wastewater treatment/re-use ratio | % | +1 | +1 | +1 | 0 | 0 | 0 | 0 | 0 | +3 | Leung 1999 |
| | Material serviceability | % | 0 | 0 | +1 | 0 | 0 | 0 | 0 | 0 | +1 | Orofino 1989; Suprenant 1990; Horvath and Hendrickson 1998; Lippiatt 1999 |
| | Material durability | % | +1 | +1 | +1 | 0 | 0 | 0 | 0 | 0 | +3 | Orofino 1989; Suprenant 1990; Horvath and Hendrickson 1998; Lippiatt 1999 |
| | Cargo packaging recycling ratio | % | 0 | 0 | +1 | 0 | +1 | +1 | 0 | 0 | +3 | Ross and Evans 2003 |
| | Generative material use ratio | % | 0 | 0 | 0 | 0 | 0 | 0 | 0 | 0 | 0 | Austin 1991; Masters 2001; Enkawa and Schvaneveldt 2001; Sawhney et al. 2002; Reddy and Jagadish 2003 |
| | Waste generating rate | % | −1 | −1 | −1 | 0 | −1 | 0 | 0 | 0 | −4 | Gavilan and Bernold 1994 |

Table 3.7 (Continued)

| Class | Environmental indicators | Unit | Potential environmental impacts ($EI_i$) | | | | | | | | | Representative references |
|---|---|---|---|---|---|---|---|---|---|---|---|---|
| | | | $EI_{i,1}$ | $EI_{i,2}$ | $EI_{i,3}$ | $EI_{i,4}$ | $EI_{i,5}$ | $EI_{i,6}$ | $EI_{i,7}$ | $EI_{i,8}$ | $\sum_{j=1}^{8} EI_{i,j}$ | |
| | Waste re-use and recycling ratio | % | +1 | 0 | +1 | 0 | 0 | 0 | 0 | 0 | +2 | Walter 1976; Gidley and Sack 1984; Rhatigan and Irwin 2001 |
| | Health and safety risk to staff | % | 0 | 0 | 0 | +1 | +1 | +1 | +1 | 0 | +4 | Morris 1976; Wong et al. 1985; Austin 1991; Sauni et al. 2001; Abdelhamid and Everett 1999; Bello et al. 2002 |
| | Required skills on staff | % | 0 | 0 | +1 | 0 | +1 | 0 | 0 | 0 | +2 | Chen 2003 |
| Time | Construction duration | day | -1 | -1 | -1 | -1 | -1 | -1 | -1 | -1 | -8 | Morris and Novak 1976 |
| | Transportation time | hour | 0 | 0 | 0 | -1 | -1 | -1 | -1 | -1 | -5 | Bernstein 1983 |
| | Construction delay risk | hour | 0 | 0 | -1 | 0 | -1 | -1 | 0 | 0 | -3 | Suprenant and Malisch 2000 |
| Cost | Construction cost | $ | -1 | -1 | -1 | -1 | -1 | -1 | -1 | -1 | -8 | Koehn 1976 |
| | Environmental control cost | $ | +1 | +1 | +1 | +1 | +1 | +1 | +1 | +1 | +8 | Parker 1998 |
| Management | ISO 14001 EMS adoption | % | +1 | +1 | +1 | +1 | +1 | +1 | +1 | +1 | +8 | Kloepfer 1997 |
| | ISO 9001 QMS adoption | % | 0 | 0 | 0 | 0 | 0 | 0 | 0 | 0 | 0 | Osuagwu 2002; Escanciano et al. 2002 |

| | Unit | $EI_{i,1}$ | $EI_{i,2}$ | $EI_{i,3}$ | $EI_{i,4}$ | $EI_{i,5}$ | $EI_{i,6}$ | $EI_{i,7}$ | $EI_{i,8}$ | References |
|---|---|---|---|---|---|---|---|---|---|---|
| Computerizations | % | +1 | +1 | +1 | +1 | +1 | +1 | +1 | +8 | Sailor 1974; Arnfalk 1999 |
| Cooperativity/Unionization risk | % | +1 | +1 | +1 | +1 | +1 | +1 | +1 | +8 | Schodek 1976 |
| Site layout suitability | % | +1 | +1 | +1 | +1 | +1 | +1 | +1 | +8 | Tatum 1978 |
| Society Public health and safety risk | % | -1 | -1 | -1 | -1 | -1 | 0 | 0 | -6 | Griffith 1994; U.S.EPA 2002a,b,c |
| Waste disposal price | $ | +1 | +1 | +1 | +1 | +1 | +1 | +1 | +8 | Austin 1991 |
| Legal involvements | % | +1 | +1 | +1 | +1 | +1 | +1 | +1 | +8 | Lavers and Shiers 2000; Grigg et al. 2001 |
| Public traffic disruptions | day | 0 | 0 | -1 | -1 | -1 | -1 | 0 | -4 | USEPA 2002a; USEPA 2002b; USEPA 2002c |
| Cargo transportation burden | ton-mile | 0 | 0 | -1 | -1 | -1 | -1 | 0 | -4 | USEPA 2002a; USEPA 2002b; USEPA 2002c |
| Nature Temperature affection risk | % | 0 | -1 | -1 | 0 | 0 | 0 | 0 | -3 | Morris 1976; Tian 2002 |
| Storm affection risk | % | 0 | 0 | -1 | -1 | -1 | 0 | 0 | -3 | Rutherford 1981; Sparks et al. 1989; Carper 1990 |
| Earthquake affection risk | % | -1 | -1 | -1 | -1 | -1 | 0 | 0 | -5 | Islam and Hashmi 1999; Maitra 1999; Rosowsky 2002 |

Notes

1 Definition of $EI_{i,j}$: $EI_{i,1}$ represents environmental impact for soil and ground contamination, $EI_{i,2}$ represents environmental impact for ground and underground water, $EI_{i,3}$ represents environmental impact for C and D waste, $EI_{i,4}$ represents environmental impact for noise and vibration, $EI_{i,5}$ represents environmental impact for dust, $EI_{i,6}$ represents environmental impact for hazardous emissions and odours, $EI_{i,7}$ represents environmental impact on wildlife and natural features, and $EI_{i,8}$ represents environmental impact on archaeology.

2 Empirical value of $EI_{i,j}$: $EI_{i,j} \in = (-1, 0, +1)$ ($-1$ represents adverse environmental impact, 0 represents indefinite environmental impact, and 1 represents favourable environmental impact).

3 Kilowatt-hour (kWh): The kilowatt-hour (symbolized kWh) is a unit of energy equivalent to one kilowatt (1 kW) of power expended for one hour (1 h) of time. An energy expenditure of 1 kWh represents 3,600,000 joules ($3.600 \times 10^6$ J).

Source: http://whatis.techtarget.com/

*Table 3.8* Environmental indicators and corresponding values of plan alternatives for the ANP model

| Classification | | Environmental indicators | Unit | EI$_i$ | Plan alternatives | | |
|---|---|---|---|---|---|---|---|
| | | | | | Plan A | Plan B | Plan C |
| 1 EA Factors | 1.1 | Fuel consumption amount (FCA) | Mjoule | −8 | 36k | 45k | 49k |
| | 1.2 | Construction duration (COD) | day | −8 | 500 | 560 | 450 |
| | 1.3 | Construction cost (COC) | M$ | −8 | 30 | 31 | 29 |
| | 1.4 | Public health and safety risk (PHS) | % | −6 | 10 | 20 | 25 |
| | 1.5 | Transportation time (TRT) | hour | −5 | 4.0k | 4.5k | 4.8k |
| | 1.6 | Earthquake affection risk (EAR) | % | −5 | 0.01 | 0.01 | 0.01 |
| | 1.7 | Electricity consumption amount (ECA) | kWh | −4 | 30k | 45k | 50k |
| | 1.8 | Water consumption amount (WCA) | ton | −4 | 3.1k | 3.8k | 4.1k |
| | 1.9 | Waste generating rate (WGR) | % | −4 | 1.2 | 3.0 | 3.5 |
| | 1.10 | Public traffic disruptions (PTD) | day | −4 | 39 | 60 | 70 |
| | 1.11 | Cargo transportation burden (CTB) | ton-mile | −4 | 450k | 500k | 550k |
| | 1.12 | Construction delay risk (CDR) | hour | −3 | 150 | 200 | 220 |
| | 1.13 | Temperature affection risk (TAR) | % | −3 | 10.0 | 8.9 | 8.7 |
| | 1.14 | Storm affection risk (SAR) | % | −3 | 2.0 | 1.8 | 1.8 |
| 2 EU Factors | 2.1 | Constructability (COB) | % | 0 | 100 | 100 | 100 |
| | 2.2 | Generative material use ratio (GMU) | % | 0 | 20 | 10 | 8 |
| | 2.3 | ISO 9001 QMS adoption (QMS) | % | 0 | 100 | 100 | 100 |
| 3 EF Factors | 3.1 | Cleaner technologies/ Automation ratio (CTA) | % | +8 | 80 | 50 | 40 |
| | 3.2 | Computerizations (PCA) | % | +8 | 80 | 80 | 80 |
| | 3.3 | Environmental control cost (ECC) | M$ | +8 | 0.8 | 0.5 | 0.5 |
| | 3.4 | ISO 14001 EMS adoption (EMS) | % | +8 | 0 | 0 | 0 |

| 3.5 | Cooperativity/Unionization risk (COP) | % | +8 | 100 | 80 | 60 |
|------|------|------|------|------|------|------|
| 3.6 | Site layout suitability (SLS) | % | +8 | 80 | 60 | 50 |
| 3.7 | Waste disposal price (WDP) | M$ | +8 | 0.10 | 0.25 | 0.29 |
| 3.8 | Legal and responsibility risk (LRR) | % | +8 | 0.10 | 0.23 | 0.32 |
| 3.9 | Health and safety risk to staff (HSR) | % | +4 | 0.10 | 0.21 | 0.28 |
| 3.10 | Wastewater treatment/ re-use ratio (WTR) | % | +3 | 90 | 50 | 40 |
| 3.11 | Material durability (MAD) | % | +3 | 100 | 80 | 80 |
| 3.12 | Cargo packaging recycling ratio (CPR) | % | +3 | 100 | 50 | 0 |
| 3.13 | Waste re-use and recycling ratio (WRR) | % | +2 | 90 | 30 | 35 |
| 3.14 | Required skills on staff (RSS) | % | +2 | 80 | 60 | 60 |
| 3.15 | Material serviceability (MAS) | % | +1 | 100 | 80 | 80 |

Notes

1  $EI_i$ value equals to $\Sigma EI_{i,j}$ (refer to Table 3.7);
2  EA Factors means environmental-adverse factors, EF Factors means environmental-friendly factors, and EU Factors means environmental-uncertainty factors;
3  The corresponding value of plan alternatives is calculated based on relative information and data in each construction plan alternative and no formulas and details have been provided for these calculations in this chapter.

ratio scales that represent relative measurements of the influence of elements that interact with respect to control criteria. The ANP is a coupling of two parts: one is a control hierarchy or network of criteria and subcriteria that control the interactions (interdependencies and feedback), another is a network of influences among the nodes and clusters. Moreover, the control hierarchy is a hierarchy of criteria and subcriteria for which priorities are derived in the usual way with respect to the goal of the system being considered. The criteria are used to compare the components of a system, and the subcriteria are used to compare the elements of a component. Steps of the ANP analysis for the environmental-conscious construction planning are laid out from Step A to Step D:

### 3.3.3.1   Step A: ANP model construction

This step aims to construct an ANP model for evaluation based on determining the control hierarchies such as benefits, costs, opportunities, and risk, as well as the corresponding criterion for comparing the components (clusters) of the system and sub-criteria for comparing the elements of the system, together with a determination of the clusters with their elements for each control criterion or subcriterion.

The env.Plan model is outlined in Figure 3.10. The decision environment consists of external environment and internal environment. In the exterior env.Plan environment, the downward arrow indicates the process of transferring data required by the ANP, the upward arrow indicates the process of feedback with evaluation results from the ANP, and the feedback process (loop) between the external environment and the internal environment indicates a circulating pipe for environmental priority evaluation of alternative construction plans. In the internal env.Plan environment, connections among four clusters and 35 nodes are modelled by two-way and looped arrows to describe the existing interdependencies. The four clusters are Plan Alternatives ($C_1$), EA Factors ($C_2$), EU Factors ($C_3$), and EF Factors ($C_4$). In correspondence with the four clusters, there are 35 nodes including 3 nodes in $C_1$ ($N_{11\sim3}$), 14 nodes in $C_2$ ($N_{21\sim14}$), 3 nodes in $C_3$ ($N_{31\sim3}$) and 15 nodes in $C_4$ ($N_{41\sim15}$). Figure 3.10 illustrates the

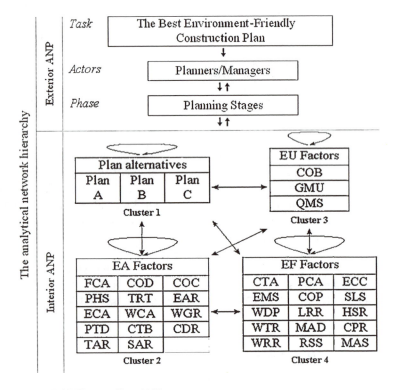

*Figure 3.10* The env.Plan ANP environment.

env.Plan model implemented using an ANP with all interior clusters and nodes, and exterior related participators.

Concerning the interdependencies between any two clusters and any two nodes, the env.Plan model structured here is a simple ANP model containing feedback and self-loops among the clusters but with no control structure because there is an implicit control criterion with respect to which all judgements (paired comparisons) are made in this model: environmental impact. For example, when comparing the cluster EA Factors ($C_2$) to cluster EF Factors ($C_4$), the latter is obviously more important for reducing negative environmental impacts, and similarly when the node comparisons are made (see Step B), relative importance of the nodes can be decided in the same way. Table 3.7 provides a list of 32 environmental indicators used in constructing the ANP model and the corresponding references from which the indicator is retrieved.

### 3.3.3.2   Step B: Paired comparisons

This step aims to perform pairwise comparisons among the clusters, as well as pairwise comparisons between nodes, as they are interdependent. On completing the pairwise comparisons, the relative importance weight (denoted as $a_{ij}$) of interdependence is determined by using a scale of pairwise judgement, where the relative importance weight is valued from 1 to 9 (Saaty 1996). The fundamental scale of pairwise judgement is given in Table 3.9. The weight of interdependence is determined by a human decision-maker who is abreast with professional experience and knowledge in the application area. In this study, it is determined subjectively as the objective of this study is mainly to demonstrate the usefulness of the ANP model in evaluating the potential environmental impact due to the execution of a construction plan.

Weights for all interdependencies for a particular construction plan are then aggregated into a series of submatrices. For example, if the cluster of plan alternatives includes Plans A, B, and C, and each of the plans is connected to nodes in

*Table 3.9* Pairwise judgements of indicator *i*

| Pairwise judgement | | 1 | 2 | 3 | 4 | 5 | 6 | 7 | 8 | 9 |
|---|---|---|---|---|---|---|---|---|---|---|
| Indicator *i* | Plan A | x | x | ✓ | x | x | x | x | x | x |
| | Plan B | x | x | x | x | ✓ | x | x | x | x |
| | Plan C | x | x | x | x | x | x | ✓ | x | x |
| Indicator $I_i$ | Indicator $I_j$ | x | x | x | x | ✓ | x | x | x | x |

Notes
1  The symbol **x** denotes item under selection for pairwise judgement, and the symbol ✓ denotes selected pairwise judgement.
2  Scale of pairwise judgement: 1 equal, 2 equally to moderately dominant, 3 moderately dominant, 4 moderately to strongly dominant, 5 strongly dominant, 6 strongly to very strongly dominant, 7 very strongly dominant, 8 very strongly to extremely dominant, 9 extremely dominant.

*Table 3.10* Formulation of supermatrix and its submatrix for env.Plan

| Supermatrix | | | | Submatrix | | | |
|---|---|---|---|---|---|---|---|

$$W = \begin{bmatrix} W_{11} & W_{12} & W_{13} & W_{14} \\ W_{21} & W_{22} & W_{23} & W_{24} \\ W_{31} & W_{32} & W_{33} & W_{34} \\ W_{41} & W_{42} & W_{43} & W_{44} \end{bmatrix}$$

Cluster : $C_1$ $C_2$ $C_3$ $C_4$
Node : $N_{1_{1\sim3}}$ $N_{2_{1\sim14}}$ $N_{3_{1\sim3}}$ $N_{4_{1\sim15}}$

$$W_{IJ} = \begin{bmatrix} w_1|_{I,J} & \cdots & w_1|_{I,J} \\ w_2|_{I,J} & \cdots & w_2|_{I,J} \\ \cdots & \cdots & \cdots \\ w_i|_{I,J} & \cdots & w_i|_{I,J} \\ \cdots & \cdots & \cdots \\ w_{N_{I_1}}|_{I,J} & \cdots & w_{N_{I_n}}|_{I,J} \end{bmatrix}$$

Notes

$I$ is the index number of rows; and $J$ is the index number of columns; both $I$ and $J$ correspond to the number of cluster and their nodes ($I, J \in (1, 2, \ldots, 35)$), $N_I$ is the total number of nodes in cluster $I$, $n$ is the total number of columns in cluster $I$. Thus a 35 × 35 supermatrix is formed.

the cluster of EF Factors, pairwise judgements of the cluster will result in relative weights of importance between each plan alternative and each EF Factor. The aggregation of the weights thus forms a 3 × 14 submatrix located at "$W_{21}$" in Table 3.10. It is necessary to note that pairwise comparisons are necessary to all connections (clusters and nodes) in the ANP model to identify the level of interdependencies which are fundamental in the ANP procedure. The series of submatrices are then aggregated into a supermatrix which is denoted as supermatrix $A$ in this study, and it will be used to derive the initial supermatrix in the later calculation in Step C.

Table 3.9 gives a general form for pairwise judgement among environmental indicators and construction plan alternatives, which is adopted in this study. For example, for the environmental indicator 1.1 Fuel consumption amount (FCA) (EA Factor 1), the pairwise judgements are as given in Table 3.9, as the fuel consumption in Plan A is the least among the three plan alternatives, whilst the fuel consumption in Plan C is the highest; in addition to this judgement in property, quantitative pairwise judgements are also made in order to define plan alternatives' priorities. After finishing a series of pairwise judgements, from environmental indicator $i$ to environmental indicator $n$, the calculation of the ANP can thus be conducted following the Step C to the Step D. Besides the pairwise judgement between an environmental indicator and a construction plan, the developed env.Plan model contains all other pairwise judgements between each of the environmental indicators (indicator $I_i$ and indicator $I_j$ in Table 3.9) and this essential initialization is set up based on the quantitative attribute of each plan alternative which has been given in Table 3.8.

### 3.3.3.3 Step C: Supermatrix calculation

This step aims to form a synthesized supermatrix to allow for the resolution of the effects of the interdependencies that exist between the elements (nodes and clusters) of the ANP model. The supermatrix of the env.Plan model is a two-dimensional partitioned matrix consisting of 16 submatrices (refer to Table 3.10).

In order to obtain useful information for construction plan selection, the calculation of the supermatrix is to be done following three substeps which transform an initial supermatrix to a weighted supermatrix, and then to a synthesized supermatrix.

At first, an initial supermatrix of the ANP model is created. The initial supermatrix consists of local priority vectors obtained from the pairwise comparisons among clusters and nodes. A local priority vector is an array of weight priorities containing a single column (denoted as $w^T = (w_1, w_2, \ldots, w_i, \ldots, w_n)$), whose components (denoted as $w_i$) are derived from a judgement comparison matrix $A$ and deduced by Equation 3.8 (Saaty 2001).

$$
w_i\big|_{I,J} = \sum_{i=1}^{I} \frac{\left( \dfrac{a_{ij}}{\sum\limits_{j=1}^{J} a_{ij}} \right)}{J}
\tag{3.8}
$$

where $w_i\big|_{I,J}$ is the weighted/derived priority of node $i$ at row $I$ and column $J$; $a_{ij}$ is a matrix value assigned to the interdependence relationship of node $i$ to node $j$. The initial supermatrix is constructed by substituting the submatrices into the supermatrix as indicated in Table 3.10. A detail of the initial supermatrix is given in Table 3.11.

After the formation of the initial supermatrix, it is transformed into a weighted supermatrix. This process involves multiplying every node in a cluster of the initial supermatrix by the weight of the cluster, which has been established by pairwise comparison among the four clusters. In the weighted supermatrix, each column is stochastic, i.e. sum of the column amounts to 1 (Saaty 2001) (refer to Table 3.12).

The last substep is to compose a limiting supermatrix, which is to raise the weighted supermatrix to powers until it converges/stabilizes, i.e. when all the columns in the supermatrix have the same values. Saaty (1996) indicated that as long as the weighted supermatrix is stochastic, a meaningful limiting result can be obtained for prediction. A limiting supermatrix can be arrived at by taking repeatedly the power of the matrix, i.e. the original weighted supermatrix, its square, its cube, etc., until the limit is attained (converges), in which case all the numbers in each row will become identical. Calculus-type algorithm is employed in the software environment of Super Decisions, designed by Bill Adams and the Creative Decision Foundation, to facilitate the formation of the limiting supermatrix, and the calculation result is listed in Table 3.12.

The formulations of supermatrices and submatrices used in the env.Plan model are illustrated in Table 3.11, and calculation results of the initial supermatrix, the weighted supermatrix, and the limiting supermatrix are given in Tables 3.11 and 3.12. As the limiting supermatrix is set up, the next step is to select a proper plan alternative using results from the limiting supermatrix.

Table 3.11 The supermatrix for the complicated env.Plan model: initial supermatrix

| Super matrix | Cluster | Node | Plan A | Plan B | Plan C | FCA | COD | COC | PHS | TRT | EAR | ECA | WCA | WGR | PTD | CTB | CDR | TAR | SAR | COB | GMU | QMS | CTA | PCA | ECC | EMS | COP | SLS | WDP | LRR | HSR | WTR | MAD | CPR | WRR | RSS | MAS |
|---|---|---|---|---|---|---|---|---|---|---|---|---|---|---|---|---|---|---|---|---|---|---|---|---|---|---|---|---|---|---|---|---|---|---|---|---|---|---|
| Initial supermatrix | Plans | Plan A | 0.33322 | 0.22554 | 0.24264 | 0.30900 | 0.16623 | 0.19420 | 0.17890 | 0.28571 | 0.19288 | 0.17003 | 0.15618 | 0.27778 | 0.19631 | 0.21764 | 0.18671 | 0.19927 | 0.19907 | 0.24883 | 0.40648 | 0.59891 | 0.65308 | 0.41261 | 0.33194 | 0.72574 | 0.25000 | 0.49339 | 0.65003 | 0.67491 | 0.71881 | 0.70905 | 0.75825 | 0.74074 | 0.74446 | 0.77858 | 0.70735 |
| | | Plan B | 0.37506 | 0.10065 | 0.08795 | 0.10945 | 0.07261 | 0.08842 | 0.20925 | 0.14286 | 0.06155 | 0.12262 | 0.18517 | 0.06703 | 0.14663 | 0.09140 | 0.07038 | 0.05455 | 0.06753 | 0.10124 | 0.33328 | 0.27491 | 0.09623 | 0.25990 | 0.52926 | 0.06205 | 0.50000 | 0.31081 | 0.17553 | 0.10124 | 0.05783 | 0.07911 | 0.15125 | 0.07407 | 0.05643 | 0.07860 | 0.12262 |
| | | Plan C | 0.10534 | 0.18970 | 0.06027 | 0.00000 | 0.09560 | 0.06693 | 0.05852 | 0.08540 | 0.08051 | 0.21142 | 0.04187 | 0.00848 | 0.16334 | 0.06290 | 0.00018 | 0.08553 | 0.12853 | 0.07170 | 0.06976 | 0.09477 | 0.09090 | 0.09182 | 0.12693 | 0.11094 | 0.09050 | 0.07435 | 0.05944 | 0.07630 | 0.10706 | 0.07990 | 0.07943 | 0.08039 | 0.08013 | 0.08565 | 0.12823 |
| | EA factors | FCA | 0.08575 | 0.09606 | 0.06923 | 0.06611 | 0.00000 | 0.07155 | 0.06768 | 0.07937 | 0.09047 | 0.09891 | 0.08121 | 0.13433 | 0.11498 | 0.14080 | 0.07879 | 0.06744 | 0.10216 | 0.13203 | 0.07099 | 0.06816 | 0.03670 | 0.08708 | 0.08669 | 0.05887 | 0.05063 | 0.07630 | 0.01076 | 0.05233 | 0.08640 | 0.07943 | 0.05556 | 0.07038 | 0.05532 | 0.08386 | 0.09811 |
| | | COD | 0.09167 | 0.06625 | 0.08715 | 0.08715 | 0.10821 | 0.09591 | 0.05006 | 0.05355 | 0.07583 | 0.12945 | 0.06155 | 0.07963 | 0.06309 | 0.07953 | 0.06175 | 0.09594 | 0.11593 | 0.07627 | 0.06549 | 0.07096 | 0.11574 | 0.06693 | 0.07471 | 0.10423 | 0.08510 | 0.09669 | 0.06815 | 0.11560 | 0.08776 | 0.07118 | 0.06663 | 0.06872 | 0.07051 | 0.07477 |
| | | COC | 0.05896 | 0.08657 | 0.07203 | 0.07203 | 0.08442 | 0.08462 | 0.07358 | 0.10135 | 0.10135 | 0.14763 | 0.06968 | 0.03993 | 0.11363 | 0.03904 | 0.07055 | 0.04312 | 0.05655 | 0.07327 | 0.08398 | 0.08817 | 0.08476 | 0.06818 | 0.08647 | 0.10097 | 0.07380 | 0.07833 | 0.06913 | 0.09519 | 0.09487 | 0.06883 | 0.07414 | 0.08194 | 0.09510 | 0.15673 | 0.09756 |
| | | PHS | 0.07590 | 0.08339 | 0.06735 | 0.06621 | 0.07140 | 0.07135 | 0.07358 | 0.00000 | 0.13044 | 0.08627 | 0.07738 | 0.06886 | 0.09180 | 0.09052 | 0.07061 | 0.08738 | 0.05887 | 0.06138 | 0.08111 | 0.08128 | 0.08128 | 0.08110 | 0.07317 | 0.05138 | 0.08110 | 0.06010 | 0.09430 | 0.09627 | 0.07280 | 0.07610 | 0.10098 | 0.08221 | 0.05656 | 0.09161 | 0.07868 |
| | | TRT | 0.06053 | 0.08280 | 0.05976 | 0.05976 | 0.16620 | 0.07150 | 0.05791 | 0.08771 | 0.00000 | 0.05205 | 0.13726 | 0.05821 | 0.05674 | 0.05821 | 0.04960 | 0.07497 | 0.09068 | 0.05076 | 0.07447 | 0.07723 | 0.07211 | 0.06421 | 0.05552 | 0.05638 | 0.07040 | 0.08161 | 0.04763 | 0.07413 | 0.08221 | 0.04882 | 0.04690 | 0.08915 | 0.07734 | 0.04626 | 0.07756 | 0.05108 |
| | | EAR | 0.05575 | 0.05925 | 0.07108 | 0.04919 | 0.07821 | 0.07640 | 0.07685 | 0.08411 | 0.07698 | 0.00000 | 0.03969 | 0.05573 | 0.05169 | 0.07886 | 0.05766 | 0.07184 | 0.13666 | 0.09464 | 0.04905 | 0.08010 | 0.06232 | 0.05490 | 0.05564 | 0.04990 | 0.04380 | 0.08970 | 0.07423 | 0.05697 | 0.04882 | 0.07608 | 0.07864 | 0.08077 | 0.08174 | 0.07102 | 0.05837 |
| | | ECA | 0.09342 | 0.05505 | 0.06545 | 0.09751 | 0.08669 | 0.06784 | 0.09035 | 0.09319 | 0.07881 | 0.07212 | 0.00000 | 0.11684 | 0.05195 | 0.07231 | 0.04960 | 0.07832 | 0.10490 | 0.06609 | 0.07301 | 0.05482 | 0.07291 | 0.05482 | 0.07291 | 0.06287 | 0.07295 | 0.05726 | 0.07828 | 0.07011 | 0.08152 | 0.07011 | 0.05727 | 0.06105 | 0.06977 |
| | | WCA | 0.03059 | 0.04846 | 0.06521 | 0.12876 | 0.06772 | 0.01280 | 0.07991 | 0.05331 | 0.07414 | 0.04960 | 0.05087 | 0.00000 | 0.05465 | 0.09608 | 0.07497 | 0.07988 | 0.07465 | 0.07285 | 0.08411 | 0.08009 | 0.08492 | 0.08894 | 0.06262 | 0.07361 | 0.08156 | 0.08557 | 0.09631 | 0.06634 | 0.05054 | 0.05054 | 0.05434 | 0.05144 | 0.03810 | 0.05216 | 0.05365 | 0.05570 |
| | | WGR | 0.06184 | 0.06204 | 0.07164 | 0.11637 | 0.06849 | 0.06940 | 0.05780 | 0.04722 | 0.09445 | 0.05766 | 0.04220 | 0.05766 | 0.00000 | 0.03781 | 0.05354 | 0.07184 | 0.13666 | 0.06493 | 0.07041 | 0.07860 | 0.04193 | 0.03352 | 0.09505 | 0.05713 | 0.05422 | 0.06941 | 0.04372 | 0.06928 | 0.07118 | 0.05815 | 0.05364 | 0.05767 | 0.04401 | 0.06551 |
| | | PTD | 0.06683 | 0.06364 | 0.09214 | 0.00003 | 0.06091 | 0.07960 | 0.10298 | 0.08735 | 0.06928 | 0.06946 | 0.07381 | 0.05863 | 0.03781 | 0.00000 | 0.07495 | 0.00000 | 0.07184 | 0.07465 | 0.08415 | 0.08009 | 0.08492 | 0.05577 | 0.05653 | 0.04703 | 0.03780 | 0.04502 | 0.07692 | 0.05753 | 0.07724 | 0.05993 | 0.06575 | 0.07065 | 0.09614 | 0.02502 | 0.05047 |
| | | CTB | 0.06287 | 0.04898 | 0.06476 | 0.05760 | 0.08005 | 0.08740 | 0.10315 | 0.04911 | 0.07998 | 0.05219 | 0.09887 | 0.06021 | 0.04510 | 0.06485 | 0.00000 | 0.06476 | 0.05492 | 0.04346 | 0.06665 | 0.06227 | 0.05238 | 0.03381 | 0.06026 | 0.09537 | 0.06865 | 0.04420 | 0.03896 | 0.05117 | 0.07766 | 0.06578 | 0.04369 | 0.06112 | 0.07515 | 0.07755 | 0.04076 |
| | | CDR | 0.07424 | 0.07131 | 0.07187 | 0.00003 | 0.06490 | 0.08447 | 0.07536 | 0.09023 | 0.03723 | 0.03565 | 0.09896 | 0.11597 | 0.05414 | 0.07815 | 0.06487 | 0.00000 | 0.04198 | 0.04644 | 0.06052 | 0.06369 | 0.05317 | 0.09109 | 0.04790 | 0.06707 | 0.05448 | 0.04909 | 0.03756 | 0.03835 | 0.05424 | 0.08642 | 0.05712 | 0.03526 | 0.04732 |
| | | TAR | 0.05830 | 0.05724 | 0.06205 | 0.00003 | 0.04736 | 0.07665 | 0.05411 | 0.07850 | 0.03151 | 0.02838 | 0.05564 | 0.04533 | 0.04533 | 0.04709 | 0.00000 | 0.04709 | 0.00000 | 0.04631 | 0.05218 | 0.06778 | 0.06778 | 0.01940 | 0.05984 | 0.04989 | 0.04974 | 0.06290 | 0.05983 | 0.04163 | 0.03501 | 0.05243 | 0.04987 | 0.03374 | 0.05361 | 0.01133 | 0.08368 |
| | | SAR | 0.14663 | 0.65481 | 0.32749 | 0.33333 | 0.25992 | 0.28488 | 0.57691 | 0.59567 | 0.64422 | 0.63484 | 0.51020 | 0.30769 | 0.21038 | 0.63699 | 0.23771 | 0.45995 | 0.21113 | 0.45996 | 0.08336 | 0.83133 | 0.21184 | 0.07812 | 0.07039 | 0.67819 | 0.43306 | 0.07558 | 0.36410 | 0.09362 | 0.28150 | 0.26837 | 0.54981 | 0.21764 | 0.39146 | 0.65065 | 0.27056 |
| | EU factors | COB | 0.19631 | 0.09534 | 0.25990 | 0.33333 | 0.33333 | 0.21739 | 0.08110 | 0.30849 | 0.42385 | 0.33381 | 0.07692 | 0.43285 | 0.04573 | 0.04383 | 0.05165 | 0.07692 | 0.22113 | 0.36806 | 0.33333 | 0.00000 | 0.16667 | 0.14242 | 0.14286 | 0.26991 | 0.14286 | 0.46644 | 0.22922 | 0.33272 | 0.28227 | 0.54355 | 0.64141 | 0.36806 | 0.09140 | 0.33013 | 0.22252 | 0.08522 |
| | | GMU | 0.65377 | 0.24986 | 0.41261 | 0.33333 | 0.32748 | 0.49773 | 0.51238 | 0.05428 | 0.37935 | 0.05091 | 0.55882 | 0.62130 | 0.61251 | 0.31899 | 0.08231 | 0.15510 | 0.31892 | 0.28628 | 0.15110 | 0.31899 | 0.08231 | 0.15510 | 0.01794 | 0.42857 | 0.21764 | 0.46644 | 0.22922 | 0.33272 | 0.28227 | 0.54355 | 0.64141 | 0.36806 | 0.09140 | 0.33013 | 0.22252 | 0.04398 |
| | | QMS | 0.06978 | 0.09930 | 0.07944 | 0.05045 | 0.12058 | 0.07396 | 0.07200 | 0.05078 | 0.06935 | 0.09011 | 0.08748 | 0.06111 | 0.08135 | 0.17483 | 0.07979 | 0.05076 | 0.06890 | 0.09972 | 0.07905 | 0.07795 | 0.00059 | 0.06772 | 0.07217 | 0.05960 | 0.03993 | 0.07416 | 0.10441 | 0.08678 | 0.07107 | 0.06924 | 0.07107 | 0.07954 | 0.05689 | 0.07970 | 0.06688 |
| | EF factors | CTA | 0.06701 | 0.04032 | 0.03982 | 0.08386 | 0.05507 | 0.02491 | 0.06818 | 0.07712 | 0.08193 | 0.04684 | 0.10862 | 0.07793 | 0.07888 | 0.08736 | 0.04049 | 0.08336 | 0.06858 | 0.06005 | 0.03884 | 0.07412 | 0.07238 | 0.05282 | 0.05302 | 0.04019 | 0.07674 | 0.06576 | 0.00000 | 0.04635 | 0.05999 | 0.08680 | 0.05477 | 0.09358 | 0.06093 | 0.08013 | 0.08480 |
| | | PCA | 0.06765 | 0.10166 | 0.06898 | 0.07213 | 0.05170 | 0.07433 | 0.06863 | 0.07873 | 0.11342 | 0.05564 | 0.08111 | 0.09453 | 0.03772 | 0.02836 | 0.06558 | 0.08111 | 0.01412 | 0.07238 | 0.05282 | 0.05302 | 0.04019 | 0.07674 | 0.05217 | 0.07679 | 0.04545 | 0.06110 | 0.00059 | 0.08513 | 0.13292 | 0.08269 | 0.08543 | 0.10780 | 0.09512 | 0.07526 | 0.01719 | 0.01402 | 0.07714 |
| | | ECC | 0.04528 | 0.07058 | 0.07249 | 0.02633 | 0.08054 | 0.06898 | 0.06505 | 0.05897 | 0.11526 | 0.03772 | 0.02836 | 0.05100 | 0.30769 | 0.21399 | 0.45995 | 0.02377 | 0.11452 | 0.09452 | 0.06109 | 0.07413 | 0.00003 | 0.06940 | 0.06940 | 0.04573 | 0.08360 | 0.08920 | 0.06605 | 0.08457 | 0.00003 | 0.07266 | 0.00268 | 0.02180 | 0.08153 |
| | | EMS | 0.07303 | 0.09716 | 0.04755 | 0.07564 | 0.07295 | 0.21739 | 0.07930 | 0.06325 | 0.06155 | 0.07934 | 0.06955 | 0.05622 | 0.06539 | 0.04285 | 0.05880 | 0.09940 | 0.08495 | 0.07807 | 0.05746 | 0.06197 | 0.05388 | 0.08136 | 0.00000 | 0.06772 | 0.07217 | 0.05960 | 0.03993 | 0.07416 | 0.11041 | 0.08678 | 0.08212 | 0.06976 | 0.07107 | 0.06924 | 0.07954 | 0.07248 |
| | | COP | 0.07125 | 0.05540 | 0.08961 | 0.07172 | 0.07120 | 0.08342 | 0.03420 | 0.07395 | 0.05428 | 0.07549 | 0.09550 | 0.07973 | 0.05547 | 0.06335 | 0.09721 | 0.08731 | 0.05295 | 0.06380 | 0.07241 | 0.10434 | 0.05613 | 0.07751 | 0.06380 | 0.07437 | 0.00000 | 0.09136 | 0.11458 | 0.07763 | 0.08908 | 0.10098 | 0.06598 | 0.06965 | 0.09144 | 0.07724 | 0.04398 |
| | | SLS | 0.07273 | 0.09930 | 0.07944 | 0.05045 | 0.04997 | 0.04530 | 0.09840 | 0.05752 | 0.05402 | 0.07200 | 0.07848 | 0.06171 | 0.07328 | 0.01160 | 0.07219 | 0.08441 | 0.09871 | 0.05290 | 0.06477 | 0.05290 | 0.14900 | 0.06885 | 0.09060 | 0.07502 | 0.08543 | 0.00001 | 0.00000 | 0.08388 | 0.10075 | 0.09265 | 0.07554 | 0.09617 | 0.08762 | 0.05689 | 0.07970 | 0.06688 |
| | | WDP | 0.06701 | 0.04023 | 0.03982 | 0.02491 | 0.02491 | 0.00618 | 0.07712 | 0.08193 | 0.01342 | 0.04684 | 0.10862 | 0.07793 | 0.07888 | 0.08736 | 0.04049 | 0.08336 | 0.06858 | 0.07412 | 0.07238 | 0.08162 | 0.07420 | 0.02738 | 0.05282 | 0.04019 | 0.07674 | 0.02293 | 0.01672 | 0.00059 | 0.04635 | 0.04136 | 0.08268 | 0.05477 | 0.09358 | 0.06093 | 0.08013 | 0.08480 |
| | | LRR | 0.05020 | 0.04022 | 0.07242 | 0.03047 | 0.06476 | 0.07738 | 0.04408 | 0.05815 | 0.06766 | 0.03772 | 0.02834 | 0.06558 | 0.05455 | 0.08147 | 0.03443 | 0.05574 | 0.06532 | 0.06669 | 0.05217 | 0.07679 | 0.04745 | 0.06410 | 0.05742 | 0.07659 | 0.08360 | 0.08920 | 0.10495 | 0.06605 | 0.06533 | 0.05881 | 0.00000 | 0.07628 | 0.04634 | 0.05996 | 0.06777 |
| | | HSR | 0.09017 | 0.07464 | 0.08631 | 0.04293 | 0.05494 | 0.07725 | 0.07469 | 0.05417 | 0.06595 | 0.06957 | 0.06551 | 0.04750 | 0.05151 | 0.08884 | 0.04930 | 0.01929 | 0.07661 | 0.06037 | 0.04745 | 0.06102 | 0.05742 | 0.07196 | 0.06533 | 0.05254 | 0.12612 | 0.00000 | 0.07628 | 0.03897 | 0.07380 | 0.07604 | 0.00000 | 0.08634 | 0.10172 | 0.07737 | 0.06165 | 0.08999 |
| | | WTR | 0.04907 | 0.03911 | 0.05085 | 0.05552 | 0.06893 | 0.04367 | 0.05882 | 0.06093 | 0.04791 | 0.05678 | 0.02535 | 0.04918 | 0.08239 | 0.05309 | 0.02290 | 0.06716 | 0.09108 | 0.05667 | 0.04848 | 0.04056 | 0.08948 | 0.07601 | 0.08936 | 0.06750 | 0.08223 | 0.05880 | 0.05484 | 0.09333 | 0.03897 | 0.00000 | 0.06445 | 0.10172 | 0.07737 | 0.00165 | 0.08999 |
| | | MAD | 0.08851 | 0.08190 | 0.07871 | 0.06837 | 0.06145 | 0.07590 | 0.06150 | 0.08811 | 0.06500 | 0.06816 | 0.08007 | 0.06791 | 0.07243 | 0.03851 | 0.03429 | 0.05593 | 0.07447 | 0.05849 | 0.06484 | 0.07637 | 0.06475 | 0.03850 | 0.09050 | 0.05727 | 0.05464 | 0.06395 | 0.04192 | 0.06795 | 0.07380 | 0.07604 | 0.00000 | 0.08634 | 0.07604 | 0.00000 | 0.05797 |
| | | CPR | 0.07507 | 0.03948 | 0.02564 | 0.07972 | 0.05084 | 0.05727 | 0.03913 | 0.05334 | 0.05722 | 0.05940 | 0.02041 | 0.05137 | 0.08164 | 0.06558 | 0.02648 | 0.06248 | 0.04896 | 0.05972 | 0.06524 | 0.08317 | 0.07762 | 0.07752 | 0.10130 | 0.05818 | 0.07710 | 0.08736 | 0.05861 | 0.04136 | 0.02714 | 0.05039 | 0.00000 | 0.05395 | 0.04600 | 0.05540 |
| | | WRR | 0.06955 | 0.06796 | 0.07446 | 0.08191 | 0.07277 | 0.07615 | 0.07065 | 0.02372 | 0.07676 | 0.08164 | 0.01870 | 0.08119 | 0.06394 | 0.04381 | 0.05278 | 0.04768 | 0.06532 | 0.07241 | 0.06496 | 0.06964 | 0.07278 | 0.05289 | 0.09420 | 0.05269 | 0.09420 | 0.07967 | 0.07062 | 0.03930 | 0.04360 | 0.06986 | 0.00000 | 0.04699 | 0.06954 |
| | | RSS | 0.05114 | 0.03328 | 0.04609 | 0.07877 | 0.04541 | 0.09009 | 0.04965 | 0.05091 | 0.05022 | 0.03130 | 0.09731 | 0.09745 | 0.05323 | 0.05240 | 0.03145 | 0.06175 | 0.04363 | 0.06011 | 0.05468 | 0.06523 | 0.04177 | 0.03005 | 0.06801 | 0.05792 | 0.06225 | 0.07039 | 0.06223 | 0.04372 | 0.06441 | 0.04255 | 0.03405 | 0.05294 | 0.00000 | 0.07032 |
| | | MAS | 0.05736 | 0.06716 | 0.06984 | 0.05677 | 0.04958 | 0.06782 | 0.09677 | 0.07297 | 0.03539 | 0.07320 | 0.06173 | 0.04675 | 0.04493 | 0.04477 | 0.03166 | 0.04493 | 0.04773 | 0.06482 | 0.05682 | 0.06311 | 0.05682 | 0.06311 | 0.04589 | 0.04560 | 0.06611 | 0.04140 | 0.03956 | 0.06092 | 0.03678 | 0.05235 | 0.03500 |

## Table 3.12 The supermatrix for the complicated env.Plan model: weighted and limiting supermatrices

| Super matrix | Cluster | | Node | Plan alternatives | | | EA factors | | | | | | | | | | | | | | EU factors | | | | | | | | | EF factors | | | | | | | | |
|---|---|---|---|---|---|---|---|---|---|---|---|---|---|---|---|---|---|---|---|---|---|---|---|---|---|---|---|---|---|---|---|---|---|---|---|---|---|---|---|
| | | | | Plan A | Plan B | Plan C | FCA | COD | COC | PHS | TRT | EAR | ECA | WCA | WGR | PTD | CTB | CDR | TAR | SAR | COB | GMU | QMS | CTA | PCA | ECC | EMS | COP | SLS | WDP | LRR | HSR | WTR | MAD | CPR | WRR | RSS | MAS |
| Weighted supermatrix | Plans | Plan A | Plan A | 0.14793 | 0.16845 | 0.16735 | 0.08330 | 0.05638 | 0.06066 | 0.07725 | 0.02736 | 0.01815 | 0.03571 | 0.03905 | 0.06794 | 0.04908 | 0.05441 | 0.04668 | 0.04336 | 0.04977 | 0.06221 | 0.06506 | 0.03155 | 0.00459 | 0.00600 | 0.00254 | 0.00389 | 0.00458 | 0.00358 | 0.00319 | 0.00040 | 0.00409 | 0.00388 | 0.00166 | 0.00339 | 0.00274 | 0.00261 | 0.00311 |
| | | Plan B | Plan B | 0.05638 | 0.06066 | 0.07725 | 0.04156 | 0.04855 | 0.04473 | 0.01463 | 0.02134 | 0.16530 | 0.16446 | 0.05441 | 0.06794 | 0.04908 | 0.03666 | 0.04629 | 0.03666 | 0.04977 | 0.19300 | 0.06506 | 0.14973 | 0.00755 | 0.00600 | 0.00254 | 0.00969 | 0.01329 | 0.00903 | 0.01190 | 0.01316 | 0.01236 | 0.02398 | 0.01388 | 0.01356 | 0.01363 | 0.01425 | 0.00311 |
| | | Plan C | Plan C | 0.06066 | 0.07725 | 0.02736 | 0.07725 | 0.02736 | 0.04855 | 0.05231 | 0.03571 | 0.16427 | 0.17274 | 0.18573 | 0.06794 | 0.04908 | 0.01760 | 0.02005 | 0.01364 | 0.03211 | 0.18573 | 0.06506 | 0.10162 | 0.01196 | 0.00600 | 0.00969 | 0.00915 | 0.00458 | 0.00569 | 0.00321 | 0.00106 | 0.00185 | 0.00145 | 0.00277 | 0.00136 | 0.00193 | 0.00144 | 0.00225 |
| EA factors | FCA | | FCA | 0.02634 | 0.02974 | 0.02007 | 0.00000 | 0.02739 | 0.01673 | 0.01463 | 0.02134 | 0.01047 | 0.03035 | 0.03035 | 0.02712 | 0.04084 | 0.01572 | 0.02005 | 0.02138 | 0.03211 | 0.02679 | 0.05735 | 0.05793 | 0.00176 | 0.00600 | 0.08008 | 0.06999 | 0.06880 | 0.04691 | 0.06021 | 0.05576 | 0.05011 | 0.06813 | 0.05949 | 0.05055 | 0.05404 | 0.07335 | 0.08090 |
| | COD | | COD | 0.01440 | 0.02401 | 0.01731 | 0.01653 | 0.00000 | 0.02739 | 0.01463 | 0.01704 | 0.01686 | 0.02875 | 0.03358 | 0.02873 | 0.02417 | 0.01970 | 0.02852 | 0.03076 | 0.01773 | 0.02554 | 0.03714 | 0.06754 | 0.04813 | 0.05450 | 0.05494 | 0.05549 | 0.03714 | 0.06638 | 0.06481 | 0.06039 | 0.05450 | 0.03505 | 0.04440 | 0.06657 | 0.05291 | 0.04936 | 0.06190 |
| | COC | | COC | 0.02742 | 0.01656 | 0.02179 | 0.02398 | 0.02705 | 0.00000 | 0.01251 | 0.01339 | 0.01896 | 0.03241 | 0.01539 | 0.01766 | 0.02971 | 0.01577 | 0.01988 | 0.01544 | 0.02398 | 0.02898 | 0.01907 | 0.01631 | 0.04393 | 0.05473 | 0.04713 | 0.07597 | 0.05369 | 0.04100 | 0.04300 | 0.02793 | 0.05369 | 0.04100 | 0.05537 | 0.04204 | 0.04335 | 0.10757 | 0.04717 |
| | PHS | | PHS | 0.04740 | 0.02164 | 0.01801 | 0.06956 | 0.02417 | 0.02116 | 0.00000 | 0.01742 | 0.00076 | 0.00000 | 0.02786 | 0.02099 | 0.00369 | 0.01742 | 0.00998 | 0.00000 | 0.00076 | 0.01764 | 0.00078 | 0.01764 | 0.05347 | 0.06370 | 0.04782 | 0.04492 | 0.04361 | 0.06005 | 0.05985 | 0.04142 | 0.04678 | 0.05170 | 0.06000 | 0.09888 | 0.06155 | 0.06000 | 0.06155 |
| | TRT | | TRT | 0.01897 | 0.00085 | 0.01684 | 0.01554 | 0.01785 | 0.01784 | 0.01840 | 0.00000 | 0.01840 | 0.00000 | 0.00000 | 0.02685 | 0.00722 | 0.02295 | 0.02263 | 0.00767 | 0.02185 | 0.01895 | 0.02032 | 0.01469 | 0.05243 | 0.03873 | 0.04616 | 0.06633 | 0.03791 | 0.05949 | 0.06073 | 0.05949 | 0.05990 | 0.06370 | 0.05186 | 0.06722 | 0.05779 | 0.04963 | 0.04963 |
| | EAR | | EAR | 0.01513 | 0.02070 | 0.01494 | 0.00001 | 0.01494 | 0.01788 | 0.01448 | 0.02194 | 0.00000 | 0.00000 | 0.02194 | 0.00000 | 0.00130 | 0.01455 | 0.01815 | 0.01355 | 0.01161 | 0.01433 | 0.01269 | 0.01931 | 0.03503 | 0.03557 | 0.00431 | 0.04442 | 0.05148 | 0.03005 | 0.04677 | 0.05186 | 0.04378 | 0.05624 | 0.04879 | 0.02919 | 0.04893 | 0.03222 | 0.03222 |
| | ECA | | ECA | 0.01940 | 0.01481 | 0.01777 | 0.01730 | 0.01955 | 0.01910 | 0.01921 | 0.02103 | 0.01925 | 0.00000 | 0.00000 | 0.01921 | 0.02103 | 0.01815 | 0.01815 | 0.03705 | 0.02274 | 0.00707 | 0.00126 | 0.02003 | 0.03447 | 0.03148 | 0.02763 | 0.05659 | 0.04683 | 0.03594 | 0.03080 | 0.04800 | 0.04961 | 0.05095 | 0.04804 | 0.04480 | 0.04480 | 0.03682 |
| | WCA | | WCA | 0.02335 | 0.01376 | 0.01636 | 0.02438 | 0.02167 | 0.01696 | 0.02259 | 0.02330 | 0.01970 | 0.01803 | 0.00000 | 0.01803 | 0.00000 | 0.01299 | 0.01502 | 0.01958 | 0.02632 | 0.01502 | 0.01825 | 0.01895 | 0.04007 | 0.05957 | 0.03458 | 0.04600 | 0.04137 | 0.03966 | 0.04602 | 0.03612 | 0.04423 | 0.05143 | 0.05160 | 0.03613 | 0.03852 | 0.04401 | 0.04401 |
| | WGR | | WGR | 0.00765 | 0.01212 | 0.01636 | 0.03219 | 0.01694 | 0.02532 | 0.01998 | 0.00383 | 0.01853 | 0.01244 | 0.01772 | 0.00000 | 0.02641 | 0.01884 | 0.02402 | 0.00874 | 0.02267 | 0.00777 | 0.01413 | 0.01589 | 0.03950 | 0.04644 | 0.05145 | 0.05407 | 0.06076 | 0.06185 | 0.03188 | 0.03430 | 0.03245 | 0.02404 | 0.03290 | 0.03385 | 0.03514 | 0.03514 |
| | PTD | | PTD | 0.01546 | 0.00151 | 0.01791 | 0.02907 | 0.01712 | 0.01195 | 0.01445 | 0.01181 | 0.02361 | 0.01442 | 0.01695 | 0.01538 | 0.00000 | 0.01339 | 0.01796 | 0.03417 | 0.03417 | 0.02366 | 0.01760 | 0.01822 | 0.02645 | 0.03377 | 0.05996 | 0.03604 | 0.03430 | 0.04379 | 0.02758 | 0.04370 | 0.05743 | 0.00491 | 0.03146 | 0.03668 | 0.03638 | 0.04113 |
| | CTB | | CTB | 0.01671 | 0.01591 | 0.02234 | 0.00001 | 0.01712 | 0.01990 | 0.02575 | 0.02184 | 0.01732 | 0.01711 | 0.00000 | 0.00000 | 0.01619 | 0.00000 | 0.01821 | 0.00874 | 0.00000 | 0.02002 | 0.15357 | 0.01087 | 0.02385 | 0.02967 | 0.02385 | 0.02284 | 0.04853 | 0.02629 | 0.04873 | 0.03756 | 0.04148 | 0.04457 | 0.06065 | 0.01578 | 0.03184 | 0.03184 |
| | CDR | | CDR | 0.01570 | 0.01224 | 0.01619 | 0.02644 | 0.02001 | 0.02185 | 0.02579 | 0.01128 | 0.01621 | 0.01305 | 0.02247 | 0.01505 | 0.00000 | 0.01373 | 0.00000 | 0.01619 | 0.01619 | 0.01647 | 0.03304 | 0.01657 | 0.03304 | 0.02133 | 0.03802 | 0.00607 | 0.04331 | 0.02789 | 0.03228 | 0.04899 | 0.01450 | 0.02757 | 0.03854 | 0.04741 | 0.00489 | 0.02571 |
| | TAR | | TAR | 0.01856 | 0.00783 | 0.01797 | 0.00001 | 0.01622 | 0.02112 | 0.01884 | 0.02256 | 0.00931 | 0.00889 | 0.02475 | 0.02824 | 0.01353 | 0.01954 | 0.01622 | 0.00000 | 0.00050 | 0.00161 | 0.01592 | 0.03354 | 0.05746 | 0.03922 | 0.04231 | 0.04231 | 0.03097 | 0.02486 | 0.02369 | 0.02420 | 0.02369 | 0.03422 | 0.05452 | 0.01603 | 0.02224 | 0.02986 |
| | SAR | | SAR | 0.01458 | 0.01431 | 0.01551 | 0.00001 | 0.00184 | 0.01691 | 0.02647 | 0.01900 | 0.00788 | 0.00770 | 0.02641 | 0.01344 | 0.01133 | 0.01177 | 0.00000 | 0.01177 | 0.00000 | 0.01158 | 0.01602 | 0.01602 | 0.01224 | 0.04276 | 0.03147 | 0.03138 | 0.03775 | 0.03147 | 0.03711 | 0.02626 | 0.03308 | 0.03146 | 0.02129 | 0.03382 | 0.00715 | 0.05379 |
| EU factors | COB | | COB | 0.03666 | 0.16370 | 0.08187 | 0.08133 | 0.06498 | 0.07122 | 0.14782 | 0.16106 | 0.15871 | 0.12755 | 0.03692 | 0.05260 | 0.15925 | 0.05943 | 0.11499 | 0.13745 | 0.00000 | 0.16667 | 0.00833 | 0.00000 | 0.00000 | 0.00970 | 0.03318 | 0.03737 | 0.02342 | 0.02409 | 0.01938 | 0.00506 | 0.01522 | 0.01451 | 0.02974 | 0.01177 | 0.02117 | 0.03519 |
| | GMU | | GMU | 0.04908 | 0.02383 | 0.06498 | 0.08133 | 0.03015 | 0.05415 | 0.02027 | 0.07712 | 0.02131 | 0.01949 | 0.08345 | 0.01923 | 0.01596 | 0.09202 | 0.05528 | 0.09202 | 0.15180 | 0.00833 | 0.16667 | 0.00833 | 0.02105 | 0.00000 | 0.04167 | 0.00444 | 0.03760 | 0.01670 | 0.01670 | 0.03375 | 0.00946 | 0.00634 | 0.00444 | 0.03737 | 0.01506 | 0.03484 |
| | QMS | | QMS | 0.16427 | 0.06246 | 0.10315 | 0.08133 | 0.08187 | 0.12443 | 0.08550 | 0.01716 | 0.01858 | 0.01716 | 0.01968 | 0.01968 | 0.03900 | 0.11387 | 0.01900 | 0.15385 | 0.03900 | 0.08333 | 0.00833 | 0.16667 | 0.08333 | 0.00970 | 0.00000 | 0.00970 | 0.01177 | 0.02523 | 0.01240 | 0.01800 | 0.01527 | 0.01991 | 0.03323 | 0.01991 | 0.00204 | 0.00461 |
| CTA factors | CTA | | CTA | 0.01744 | 0.03395 | 0.02437 | 0.03015 | 0.00189 | 0.01769 | 0.01221 | 0.04892 | 0.01606 | 0.15871 | 0.00734 | 0.02253 | 0.01287 | 0.01291 | 0.00208 | 0.01291 | 0.04371 | 0.00000 | 0.02105 | 0.00000 | 0.02105 | 0.00000 | 0.02886 | 0.02516 | 0.02377 | 0.02844 | 0.03498 | 0.02516 | 0.02241 | 0.03498 | 0.02150 | 0.02150 | 0.02655 | 0.01612 |
| | PCA | | PCA | 0.01691 | 0.02542 | 0.01724 | 0.01803 | 0.02629 | 0.01858 | 0.01716 | 0.01968 | 0.01968 | 0.01949 | 0.02028 | 0.01613 | 0.02028 | 0.01686 | 0.00722 | 0.00515 | 0.01513 | 0.02288 | 0.02044 | 0.00084 | 0.03704 | 0.02422 | 0.00000 | 0.02287 | 0.01710 | 0.02985 | 0.02526 | 0.03944 | 0.00454 | 0.02535 | 0.03199 | 0.02823 | 0.02233 | 0.02289 |
| ECC factors | ECC | | ECC | 0.01120 | 0.01764 | 0.01620 | 0.01663 | 0.02013 | 0.01724 | 0.00521 | 0.01474 | 0.02881 | 0.01732 | 0.02035 | 0.00713 | 0.01787 | 0.01760 | 0.03450 | 0.01481 | 0.01686 | 0.00000 | 0.02995 | 0.02413 | 0.02995 | 0.03704 | 0.00000 | 0.02214 | 0.02805 | 0.01813 | 0.02206 | 0.02059 | 0.01960 | 0.02509 | 0.02968 | 0.02155 | 0.03047 | 0.03032 |
| | EMS | | EMS | 0.01826 | 0.02429 | 0.01189 | 0.00191 | 0.01911 | 0.01539 | 0.01982 | 0.00581 | 0.01649 | 0.01649 | 0.01984 | 0.01738 | 0.01406 | 0.01635 | 0.01649 | 0.00406 | 0.01738 | 0.01436 | 0.01839 | 0.01952 | 0.01436 | 0.01599 | 0.02414 | 0.00000 | 0.02207 | 0.00000 | 0.02303 | 0.02663 | 0.01958 | 0.00067 | 0.02575 | 0.02109 | 0.02055 | 0.02151 |
| | COP | | COP | 0.01810 | 0.01385 | 0.01220 | 0.00768 | 0.01780 | 0.02085 | 0.01357 | 0.00897 | 0.02388 | 0.02411 | 0.01487 | 0.00276 | 0.00887 | 0.02388 | 0.02430 | 0.02088 | 0.03214 | 0.01804 | 0.02207 | 0.02037 | 0.01804 | 0.00892 | 0.02018 | 0.00892 | 0.00000 | 0.02271 | 0.03400 | 0.02303 | 0.02663 | 0.02993 | 0.01958 | 0.00067 | 0.02713 | 0.01305 |
| | SLS | | SLS | 0.01810 | 0.02482 | 0.00986 | 0.01261 | 0.01249 | 0.00133 | 0.02460 | 0.01431 | 0.00135 | 0.01962 | 0.01143 | 0.01832 | 0.00290 | 0.01143 | 0.02110 | 0.02469 | 0.02075 | 0.02043 | 0.02226 | 0.02535 | 0.01669 | 0.02688 | 0.02226 | 0.02135 | 0.01632 | 0.00000 | 0.02489 | 0.02990 | 0.02749 | 0.02242 | 0.02854 | 0.02600 | 0.01688 | 0.02044 |
| EF factors | WDP | | WDP | 0.06750 | 0.01008 | 0.00995 | 0.02097 | 0.01377 | 0.01705 | 0.01948 | 0.01928 | 0.00117 | 0.02715 | 0.01948 | 0.00177 | 0.02715 | 0.01948 | 0.00512 | 0.02064 | 0.01714 | 0.01501 | 0.01786 | 0.01573 | 0.01786 | 0.01567 | 0.01573 | 0.01193 | 0.03277 | 0.01951 | 0.00000 | 0.01375 | 0.01786 | 0.02576 | 0.01685 | 0.02777 | 0.01808 | 0.02516 |
| | LRR | | LRR | 0.01270 | 0.00983 | 0.01271 | 0.00762 | 0.01469 | 0.01188 | 0.00723 | 0.00171 | 0.01995 | 0.02103 | 0.00117 | 0.00943 | 0.01634 | 0.00551 | 0.00865 | 0.01736 | 0.02647 | 0.02647 | 0.02505 | 0.01491 | 0.00000 | 0.00314 | 0.02481 | 0.00647 | 0.02505 | 0.00491 | 0.00000 | 0.01960 | 0.00331 | 0.01627 | 0.01736 | 0.00736 | 0.01829 | 0.02670 |
| | HSR | | HSR | 0.02213 | 0.00955 | 0.00968 | 0.00709 | 0.01538 | 0.00898 | 0.00931 | 0.00354 | 0.01538 | 0.02203 | 0.01625 | 0.00629 | 0.01188 | 0.01188 | 0.00629 | 0.01739 | 0.01698 | 0.01811 | 0.00966 | 0.01186 | 0.00525 | 0.01436 | 0.02279 | 0.02806 | 0.00000 | 0.02031 | 0.01559 | 0.03743 | 0.00000 | 0.02263 | 0.02298 | 0.01375 | 0.02777 | 0.02516 |
| | WTR | | WTR | 0.01277 | 0.00093 | 0.01271 | 0.01388 | 0.00723 | 0.00092 | 0.01471 | 0.02715 | 0.02190 | 0.01986 | 0.00198 | 0.00614 | 0.00230 | 0.00000 | 0.01337 | 0.00692 | 0.00063 | 0.01392 | 0.01212 | 0.00114 | 0.02665 | 0.02255 | 0.02003 | 0.02440 | 0.01745 | 0.01627 | 0.01156 | 0.00000 | 0.01786 | 0.01736 | 0.00442 | 0.00000 | 0.02296 | 0.01744 |
| | MAD | | MAD | 0.02213 | 0.00955 | 0.00968 | 0.02048 | 0.00188 | 0.01904 | 0.02030 | 0.00203 | 0.01625 | 0.01704 | 0.02018 | 0.01698 | 0.01811 | 0.02030 | 0.00857 | 0.00963 | 0.01811 | 0.01907 | 0.00462 | 0.01621 | 0.00621 | 0.01907 | 0.01921 | 0.01142 | 0.02685 | 0.01898 | 0.01244 | 0.02331 | 0.02190 | 0.00000 | 0.02256 | 0.00000 | 0.02562 | 0.02978 |
| | CPR | | CPR | 0.01877 | 0.00987 | 0.00564 | 0.00993 | 0.01271 | 0.01432 | 0.00979 | 0.01334 | 0.01430 | 0.01485 | 0.01493 | 0.00771 | 0.01324 | 0.00572 | 0.00857 | 0.01362 | 0.01240 | 0.00641 | 0.01493 | 0.00641 | 0.01493 | 0.02003 | 0.03200 | 0.03006 | 0.01726 | 0.02592 | 0.01739 | 0.00127 | 0.01495 | 0.00000 | 0.01495 | 0.00000 | 0.01601 | 0.01644 |
| | WRR | | WRR | 0.01739 | 0.01699 | 0.01862 | 0.02048 | 0.01188 | 0.01904 | 0.02030 | 0.01917 | 0.01917 | 0.01493 | 0.02030 | 0.02717 | 0.02030 | 0.01320 | 0.01192 | 0.01493 | 0.01493 | 0.01810 | 0.01624 | 0.01741 | 0.01810 | 0.01624 | 0.01741 | 0.02160 | 0.01570 | 0.02795 | 0.01564 | 0.01622 | 0.01998 | 0.01294 | 0.01294 | 0.00000 | 0.01395 | 0.02067 |
| | RSS | | RSS | 0.02780 | 0.00820 | 0.01152 | 0.01969 | 0.01135 | 0.02252 | 0.01241 | 0.01273 | 0.01310 | 0.00782 | 0.02433 | 0.02436 | 0.01310 | 0.00864 | 0.01627 | 0.01310 | 0.00864 | 0.01503 | 0.01367 | 0.01503 | 0.01367 | 0.01936 | 0.01818 | 0.00892 | 0.02018 | 0.01719 | 0.00667 | 0.02711 | 0.00067 | 0.01025 | 0.01371 | 0.00010 | 0.00000 | 0.01039 |
| | MAS | | MAS | 0.01439 | 0.01679 | 0.01746 | 0.01419 | 0.01215 | 0.01696 | 0.02419 | 0.01824 | 0.00885 | 0.01830 | 0.00403 | 0.01215 | 0.01830 | 0.00792 | 0.01123 | 0.01119 | 0.01119 | 0.00968 | 0.01184 | 0.01989 | 0.01184 | 0.01923 | 0.01816 | 0.02248 | 0.00873 | 0.01362 | 0.01353 | 0.01962 | 0.01229 | 0.01174 | 0.01808 | 0.01091 | 0.01554 | 0.01039 |
| Limiting supermatrix | | | $N_i$ | 0.1123 | 0.04149 | 0.03543 | 0.03296 | 0.03001 | 0.0305 | 0.02777 | 0.02779 | 0.02271 | 0.02222 | 0.02280 | 0.02155 | 0.02271 | 0.02268 | 0.02091 | 0.02048 | 0.01879 | 0.08042 | 0.04520 | 0.07300 | 0.02196 | 0.0201 | 0.0199 | 0.01859 | 0.01942 | 0.01899 | 0.01658 | 0.01562 | 0.01725 | 0.01513 | 0.01749 | 0.01546 | 0.0172 | 0.01416 | 0.01428 |

### Note

$N_i$ stands for any of the 35 nodes involved in the four clusters including Plan alternatives, EA factors, EU factors, and EF factors.

### 3.3.3.4 Step D: Selection

This step aims to select the best construction plan based on the computation results of the limiting supermatrix of the ANP model. Main results of the ANP model computations are the overall priorities of construction plans obtained by synthesizing the priorities of individual construction plans against different environmental indicators. The selection of the best construction plan, which has the highest environmental priority, can be done using a limiting priority weight, which is defined in Equation 3.9.

$$W_i = w_{C_{\text{Plan}},i} / w_{C_{\text{Plan}}} = w_{C_{\text{Plan}},i} / (w_{C_{\text{Plan}},1} + \cdots + w_{C_{\text{Plan}},n}) \tag{3.9}$$

where $W_i$ is the synthesized priority weight of plan alternative $i (i = 1, \ldots, n)$ ($n$ is the total number of plan alternatives, $n = 3$ in this study), and $w_{C_{\text{Plan}},i}$ is the limited weight of plan alternative $i$ in the limiting supermatrix. Because the $w_{C_{\text{Plan}},i}$ is transformed from pairwise judgements conducted in Step B, it is reasonable to regard it as the priority of the plan alternative $i$ and thus to be used in Equation 3.9. According to the computation results in the limiting supermatrix in Table 3.12, $w_{C_{\text{Plan}},i} = (0.11231, 0.04149, 0.03543)$, so $W_i = (0.59351, 0.21926, 0.18723)$; as a result, the best environmental-conscious construction plan is Plan A.

In addition to the complicated env.Plan model developed in Figure 3.10, another ANP model, called simplified env.Plan model for alternative construction plan selection, was developed with 15 nodes selected from the total 35 nodes of the complicated env.Plan model in Figure 3.10. In order to decrease the number of elements in a supermatrix of the simplified env.Plan model, similar subcomponents of EF Factors are combined, including a combination of subcomponents 3.1 and 3.2 for environment-friendly construction and management technology (Technology) and a combination of subcomponents 3.3 and 3.4 for environmental control cost (ECC). Finally, the nodes for the simplified env.Plan model include FCA, COD, and COC in EA Factors cluster; COB, GMU, and QMS in the EU Factors cluster; CTA + PCA, ECC + EMS, COP, SLS, WDP, and LRR in the EF Factors cluster; and Plan A, Plan B, and Plan C in the Plan Alternatives cluster. The rule for selecting nodes for the EA Factors cluster and the EF Factors cluster of the simplified env.Plan model is whether the absolute value of $EI$ is 8. In other words, all factors with a EI value of $-8$ go to EA cluster, and all factors with a EI of $+8$ go to EF cluster; all other factors are therefore ignored for the simplified env.Plan model. According to the computation results in the synthesized supermatrix for the simplified env.Plan model, $w_{C_{\text{Plan}},i} = (0.110243, 0.036108, 0.042977)$, so $W_i = (0.58229, 0.19072, 0.22700)$, so Plan A is also selected.

Interestingly, both complicated env.Plan model and simplified env.Plan model led to the same conclusion that Plan A is the best environmental-conscious construction plan. Besides the selected plan, it is also noticed that priority queues of these plan alternatives are also equivalent (refer to Table 3.13). Considering the load of performing pairwise comparisons on the clusters and nodes would be

*Table 3.13* A comparison between the two env.Plan models using priority weight

| ANP model | No. of nodes | Synthesized priority weight $W_i$ | | | Selected plan |
|---|---|---|---|---|---|
| | | Plan A | Plan B | Plan C | |
| Simplified model | 15 | 0.58229 | 0.19072 | 0.22700 | Plan A |
| Complicated model | 35 | 0.59351 | 0.21926 | 0.18723 | Plan A |

multiplied many times in a complicated env.Plan model, the simplified env.Plan model appears to be more practicable and efficient.

According to the attributes of plan alternatives listed in Table 3.8, the comparison results using $W_i$ also imply that the most preferable plan for environmental-conscious construction is the plan that regulates the construction practice with least consumption on fuel and water, a lowest ratio of wastage, and a maximum ratio of recycle and re-use on materials and packaging, etc. This indicates the env.Plan method can provide a quite reasonable comparison result for environmental-conscious construction and thus can be applied into construction practice.

### 3.3.4  Recommendations

In summary, in order to apply the env.Plan model in practice, the following steps are recommended:

1.  selection of an ANP model between the simplified env.Plan model and the complicated env.Plan model;
2.  original assessment of plan alternatives based on all environmental indicators, using Table 3.8;
3.  pairwise comparisons among all environmental indicators using Table 3.9;
4.  supermatrix calculation following the three substeps to transform an initial supermatrix to a limiting supermatrix with reference to Tables 3.11 and 3.12;
5.  calculation of limiting priority weight of each plan alternative using limiting supermatrix and decision-making on plan alternatives using Table 3.13;
6.  if none of the plan alternatives meets environmental requirements, adjustments to the plans are needed and re-evaluation of the plans by repeating the procedure from step 2.

## 3.4  An ANP model for demolition planning[1]

### 3.4.1  Background

Demolition is an activity to disassemble and destroy a building or parts of a building for reconstruction or renovation. In general, the demolition procedure can be

1 A collaborative research with Professor Chimay Anumba and Dr Arham Abdullah.

divided into four main stages (BSI 2000; Abdullah and Anumba 2002a,b): tendering stage, pre-demolition stage, actual demolition stage, and post-demolition stage. Because demolition is regarded as a reversed process of construction (Miller 1999) demolition contractors usually use similar management methods in their projects. For example, demolition planning, just like construction planning, is also conducted at the tendering stage. Moreover, the technical aspects considered in construction planning, such as techniques, resources, duration, and site layout (Hendrickson and Au 2000), are involved in demolition planning also.

In order to select the best demolition plan for a demolition project, Kasai (1998) suggested that there are 8 criteria including structural form of the building, location of the building, permitted level of nuisance, scope of demolition, use of building, safety, and demolition period, etc. On the other hand, Abdullah and Anumba (2002a,b) developed an AHP model with six criteria: structure characteristics, site conditions, demolition cost, past experience, time, and re-use and recycling. And their case studies indicated that the AHP model could effectively help demolition contractors to select appropriate techniques for their demolition projects. Moreover, both of the two research works concluded that the decision-makers of demolition planning have to keep in mind that health and safety are the main concerns in the selection process, and the selection of the most appropriate demolition technique could be subject to a unique combination of these criteria.

Previous research has proven the usefulness of AHP in selecting the most appropriate demolition technique for any given demolition project during the planning stage. However, the calculation results in an AHP model where interrelationships among clusters are ignored may be different if the interrelationships among the clusters are considered. For example, besides the influence on the final decision on the selection of best demolition technique, the structure characteristics can also influence other clusters in the AHP model, such as cost, time, and re-use and recycling (Abdullah and Anumba 2002a,b). In fact, this problem can be solved by using ANP, which is a natural generalization and extension of the AHP that allows feedback and dependence among decision elements and clusters of elements (Saaty 1996). In this section, the authors introduce an ANP model (named DEMAN) using the same criteria and subcriteria, which are transplanted from the AHP model (named DEMAP) developed in previous research works by Abdullah and Anumba (2002a,b) and Anumba *et al.* (2003). And a comparison of the calculation results between DEMAP and DEMAN is then made.

### 3.4.2 Statement of problem

#### 3.4.2.1 Demolition planning

Demolition planning is an essential and necessary activity in the management and execution of demolition projects. It is usually conducted with several technical aspects corresponding to what are normally involved in construction planning, such as the choice of demolition techniques and plans. As an essential and challenging task, demolition planning has to not only strive to meet common concerns such as time, cost, and safety requirement, but also explore possible measures to minimize adverse environmental impacts of the demolition projects at the outset.

### 3.4.2.2 Evaluation criteria

In order to evaluate the advantage in different demolition plan alternatives, the authors use the same evaluation criteria that have been developed for best demolition technique selection in previous researches (Abudayyeh *et al.* 1998; Fesseha 1999; Abdullah and Anumba 2002a,b; Anumba *et al.* 2003), as the contents of the demolition technique evaluation and the demolition plan evaluation are similar. Thus, there are 6 main criteria and a total of 17 sub-criteria transplanted (see Section 3.4.3.1) for the selection of best demolition plan, and all these evaluation criteria are described in Table 3.14 (Abdullah and Anumba 2002a,b).

### 3.4.2.3 A demonstration project

In order to compare the calculation results from the AHP and the ANP, the authors transplant criteria from previous studies into one demonstration demolition project. Table 3.14 illustrates characteristics of three demolition plan alternatives in the demonstration project based on the criteria. The three demolition plan alternatives are the plan using progressive demolition method (DTPM plan), the plan using deliberate collapse mechanism method (DTAM plan), and the plan using deconstruction method (DTDM). Regarding the criteria adopted, this comparative study does not include characteristics other than these 17 variables (refer to Table 3.14), which are also potential criteria for the evaluation of demolition plans.

## 3.4.3 Methodology

The methodology adopted in this research is the transplantation of evaluation criteria from previous studies on the selection of best demolition technique, the construction of an ANP model using the evaluation criteria, and comparison between the calculation results from the proposed MCDM models.

### 3.4.3.1 Transplantation of evaluation criteria

As mentioned in Section 3.4.2, the evaluation criteria developed for selecting the best demolition technique consist of 6 main criteria and 17 sub-criteria from previous research (Abdullah and Anumba 2002a,b; Anumba *et al.* 2003). The transplantation of these evaluation criteria from the selection of demolition

*Table 3.14* Indicators and their corresponding values of plan alternatives for the AHP/ANP model

| Classification (cluster) | Technique indicator (Node) | Unit | Plan alternatives | | |
| --- | --- | --- | --- | --- | --- |
| | | | DTPM[a] | DTAM[a] | DTDM[a] |
| Structure characteristics (SCH) | Height (SCHH) | Storey | 12 | 12 | 12 |
| | Type (SCHT) | – | PRCS[b] | PRCS[b] | PRCS[b] |
| | Stability (SCHS) | – | Stable | Stable | Stable |
| | Degree/Extent of demolition (SCHD) | – | Full | Full | Full |
| | Use of the structure (SCHU) | – | Housing | Housing | Housing |
| Site conditions (SCD) | Health and safety for the person on/off site (SCDH) | – | Medium | Low | High |
| | Acceptable level of noise (SCDN) | dB(A) | 70–74 | 70–74 | 70–74 |
| | Proximity of the adjacent structures (SCDP) | Meter | 50 | 50 | 50 |
| | Site accessibility (SCDA) | – | Accessible | Accessible | Accessible |
| Cost (DTC) | Machinery (DTCE) (Lump sum) | £ | 50,000 | 30,000 | 50,000 |
| | Manpower (DTCW) (Lump sum) | £ | 65,000 | 70,000 | 75,000 |
| Past experiences (PED) | Familiarity with a specified technique (PEDS) | – | Familiar | Familiar | Unfamiliar |
| | Availability of plant and equipment (PEDP) | – | Available | Available | Available |
| | Availability of expertise (PEDE) | – | Available | Available | Available |
| Re-use and recycling (DTR) | Level of re-use and recycling (DTRL) | – | Moderate | Moderate | Moderate |
| Time (DTT) | Site preparation (DTTP) | Month | 3 | 3 | 3 |
| | Actual demolition (DTTD) | Month | 3 | 3 | 3 |

Notes
a DTPM acts as progressive demolition plan, DTAM acts as deliberate collapse mechanism plan, and DTDM acts as deconstruction plan.
b PRCS acts as precast reinforced concrete structure.

techniques to the selection of demolition plan requires verification of transplantation alternatives and assumptions on account of the relative uniformity and difference between the selection of demolition techniques and the selection of demolition plans. In this section, after a comparative study of the two kinds of selection, the authors finally chose an intact transplantation of the evaluation criteria from the developed model for selecting the best demolition technique.

### 3.4.3.2 Selection of ANP

The ANP is more powerful in modelling complex decision environments than the AHP because it can be used to model very sophisticated decisions involving a variety of interactions and dependencies (Meade and Sarkis 1999; Saaty 1999). The ANP is a natural generalization and extension of the AHP that allows feedback and dependence among decision elements and clusters of elements. It is also a general theory of relative measurement used to derive composite priority ratio scales from individual ratio scales that represent relative measurements of the influence of elements that interact with respect to control criteria (Saaty 1996, 1999). All these advantages are embodied in several examples of applications of the ANP (Srisoepardani 1996). For example, Meade and Sarkis (1999) applied the ANP to the strategic evaluations of environmental practices and programmes in both manufacturing and business to help analyse various project-, technological- or business-decision alternatives. Therefore, Saaty (1996) recommended the ANP be used for cases where the most thorough and systematic analysis of influences needs to be made.

### 3.4.4 DEMAN model

#### 3.4.4.1 Model construction

This section aims to construct an ANP model for selecting the best demolition plan based on the determined control hierarchy components used in the DEMAP model: structure characteristics, site condition, costs, past experience, environmental protection, and time. Meanwhile, the corresponding criteria for comparing these components (clusters) and sub-criteria for comparing the elements (nodes) of these components of the DEMAP system will be employed to compare the DEMAN model with the DEMAP model. According to the definition given by Saaty (1996), a cluster is connected to another cluster when at least one element in it is connected to at least one element in another cluster. Moreover, a determination of the clusters with their nodes for each control criterion or sub-criterion will also be done for the final comparison. The DEMAN model is outlined in Figure 3.11.

The DEMAN environment includes exterior environment and interior environment. In the exterior DEMAN environment, the downward arrow indicates the

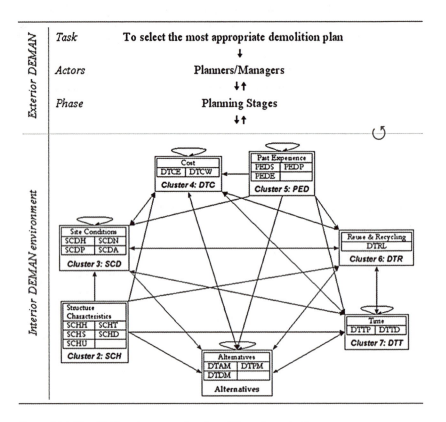

*Figure 3.11* The ANP environment for demolition plan selection.

process of transferring data required by the DEMAN, while the upward arrow indicates the process of feedback with evaluation results from the DEMAN. On the other hand, the feedback process (loop) (denoted by ↺) between the exterior environment and the interior environment indicates a circulating process for the selection of alternative demolition technique plans. In the interior DEMAN environment, connections among 7 clusters and 20 nodes are modelled by two-way and looped arrows to describe the existing interdependencies. The 7 clusters are demolition technique plan alternatives (DTA), structure characteristics (SCH), site conditions (SCD), cost (DTC), past experience (PED), re-use and recycling (DTR), and time (DTT). In correspondence with these 7 clusters, there are 20 nodes: 3 nodes in DTA, 5 nodes in SCH, 4 nodes in SCD, 2 nodes in DTC, 3 nodes in PED, 1 node in DTR, and 2 nodes in DTT. All these clusters and nodes are also described in Table 3.14. Figure 3.11 illustrates the DEMAN model implemented using an ANP with all interior clusters and nodes, and exterior-related participators.

### 3.4.4.2 Pairwise comparisons

Concerning the interdependencies between any two clusters and any two nodes, the pairwise comparisons between clusters, as well as pairwise comparisons between nodes are performed as they are interdependent. On completing the pairwise comparisons, the relative importance weight (denoted as $a_{ij}$) of interdependence is determined by using a scale of pairwise judgement, where the relative importance weight is valued from 1 to 9 (Saaty 1996). The fundamental scale of pairwise judgement is given in Table 3.15.

Table 3.15 gives a general form for pairwise judgements between any two clusters and between any two nodes in the DEMAN model. The relative importance weight of interdependence is determined manually to reflect professional experience and knowledge in the application area. In this study, the authors determine it, as the objective of this study is mainly to demonstrate the usefulness of the ANP model in selecting the best demolition plan. For example, the

*Table 3.15* Pairwise judgements between clusters/nodes in the DEMAN model

| Clusters/Nodes | | Pairwise judgements | | | | | | | | | |
|---|---|---|---|---|---|---|---|---|---|---|---|
| | | $\pm 1$ | $\pm 2$ | $\pm 3$ | $\pm 4$ | $\pm 5$ | $\pm 6$ | $\pm 7$ | $\pm 8$ | $\pm 9$ | Scales of pairwise judgements (Saaty, 1996) |
| Cluster $I$ | Cluster $J$ | x | x | x | x | x | ✓ | x | x | x | 1 = Equal, |
| Node $I_i$ | Node $J_j$ | x | x | x | x | x | ✓ | x | x | x | 2 = Equally to moderately dominant, |
| | | | | | | | | | | | 3 = Moderately dominant, |
| | | | | | | | | | | | 4 = Moderately to strongly dominant, |
| | | | | | | | | | | | 5 = Strongly dominant, |
| | | | | | | | | | | | 6 = Strongly to very strongly dominant, |
| | | | | | | | | | | | 7 = Very strongly dominant, |
| | | | | | | | | | | | 8 = Very strongly to extremely dominant, |
| | | | | | | | | | | | 9 = Extremely dominant. |

Notes
1 The symbol **x** denotes item under selection for pairwise judgement, and the symbol ✓ denotes selected pairwise judgement.
2 $I$ and $J$ denote the number of clusters, whilst $i$ and $j$ denote the total number of nodes.
3 The symbol $\pm$ denotes importance initiative between compared nodes or clusters.

relative importance weights among cluster 2 to 7 are the same as what they are in the DEMAN model (refer to Table 3.15), and the relative importance weights between cluster 1 and any one of the other six clusters are set as 1. On the other hand, the relative importance weights between any two nodes, which have a potential interdependence relationship, are set up based on the quantitative or qualitative attribute of each node in the demolition plan which has been given in Table 3.14. As a result, all pairwise comparisons between any two clusters and between any two nodes are defined according to their potential relationship based on the given scale of pairwise judgements.

Weights for all interdependencies of a particular demolition plan are then aggregated into a series of submatrices. For example, provided that the cluster of plan alternatives (DTA) includes DTAM, DTPM and DTDM, and each of these plan alternatives is connected to nodes in the cluster of cost (DTC), pairwise judgements of the cluster result in relative weights of importance between each plan alternative and each cost factor. The aggregation of the weights thus forms a $3 \times 2$ submatrix located at $W_{41}$ in Table 3.16. It is necessary to note that pairwise comparisons are necessary to all potential connections between clusters and between nodes in the DEMAN model to identify the level of interdependencies which are fundamental in the ANP procedure. The series of submatrices are then aggregated into a supermatrix, which is denoted as supermatrix $A$ in this study, and it will be used to derive the initial supermatrix in later calculations.

### 3.4.4.3  Supermatrix calculation

The supermatrix of the DEMAN system is a two-dimensional partitioned matrix consisting of 49 submatrices (refer to Table 3.16). The calculation of supermatrix aims to form a synthesized supermatrix to allow for the resolution of the effects of the interdependencies that exist between the nodes and the clusters of the ANP model. In order to obtain useful information for demolition plan selection, the

*Table 3.16* Formulation of supermatrix and its submatrix for the DEMAN

| Supermatrix | Submatrix |
|---|---|
| $W = \begin{bmatrix} W_{11} & W_{12} & \cdots & W_{17} \\ W_{21} & W_{22} & \cdots & W_{27} \\ \cdots & \cdots & \cdots & \cdots \\ W_{71} & W_{72} & \cdots & W_{77} \end{bmatrix}$ <br> Cluster: $C_1$ $C_2$ $\cdots$ $C_7$ <br> Node: $N_{1_{1 \sim 3}} N_{2_{15}} \cdots N_{7_{1 \sim 2}}$ | $W_{ij} = \begin{bmatrix} w_1 \big|_{I,J} & \cdots & w_1 \big|_{I,J} \\ w_2 \big|_{I,J} & \cdots & w_2 \big|_{I,J} \\ \cdots & & \cdots \\ w_i \big|_{I,J} & \cdots & w_i \big|_{I,J} \\ \cdots & & \cdots \\ w_{N_{I_1}} \big|_{I,J} & \cdots & w_{N_{I_n}} \big|_{I,J} \end{bmatrix}$ |

Notes

$I$ is the index number of rows; and $J$ is the index number of columns; both $I$ and $J$ correspond to the number of clusters and their nodes ($I, J \in (1, 2, \dots, 20)$), $N_I$ is the total number of nodes in cluster $I$, $n$ is the total number of columns in cluster $I$. Thus a $20 \times 20$ supermatrix is formed.

calculation of supermatrix is to be done following three steps which transform an initial supermatrix to a weighted supermatrix, and then to a synthesized supermatrix.

At first, an initial supermatrix of the DEMAN model is created. The initial supermatrix consists of local priority vectors obtained from the pairwise comparisons among clusters and nodes. A local priority vector is an array of weight priorities containing a single column (denoted as $w^T = (w_1, w_2, \ldots, w_i, \ldots, w_n)$), whose components (denoted as $w_i$) are derived from a judgement comparison matrix $A$ and deduced by Equation 3.8 in Section 3.3.3.3. The initial supermatrix is constructed by substituting the submatrices into the supermatrix as indicated in Table 3.16. A detail of the initial supermatrix is given in Table 3.17.

After the formation of the initial supermatrix, it is transformed into a weighted supermatrix by multiplying every node in a cluster of the initial supermatrix by the weight of the cluster, which has been established by pairwise comparison among the seven clusters. In the weighted supermatrix, each column is stochastic, i.e. sum of a column amounts to 1 (Saaty 2001) (refer to Table 3.17).

The last step is to compose a limiting supermatrix, which is to raise the weighted supermatrix to powers until it converges/stabilizes, i.e. when all the columns in the supermatrix have the same values. Saaty (1996) indicated that as long as the weighted supermatrix is stochastic, a meaningful limiting result could be obtained for prediction. The approach to arrive at a limiting supermatrix is by taking repeatedly the power of the matrix, i.e. the original weighted supermatrix, its square, its cube, etc., until the limit is attained (converges), in which case all the numbers in each row will become identical. Calculus-type algorithm is employed in the software environment of Super Decisions, designed by Bill Adams and the Creative Decision Foundation, to facilitate the formation of the limiting supermatrix, and the calculation result is listed in Table 3.17.

### 3.4.4.4 Demolition plan selection

The selection aims to choose the best demolition plan based on the computation results of the limiting supermatrix of the ANP model. Main results of the ANP model computations are the overall priorities of the alternatives obtained by synthesizing the priorities of individual demolition plans against different technique indicators (nodes). The selection of the best demolition plan, which has the highest priority for technological advantage, can be done using a limiting priority weight, which is defined in Equation 3.9 in Section 3.3.3.4. For the specified decision-making problem, $W_i$ is the synthesized priority weight of plan alternative $i(i = 1, \ldots, n)$ ($n$ is the total number of demolition plan alternatives, $n = 3$ in this study), and $w_{C_{Plan},i}$ is the limited weight of demolition plan alternative $i$ in the limiting supermatrix. Because the $w_{C_{Plan},i}$ is transformed from pairwise judgements, it is reasonable to regard it as the priority of the plan alternative $i$ and thus to be used in Equation 3.9. According to the computation results in the

Table 3.17 The supermatrix for the DEMAN

### Initial supermatrix

| Nodes | DTAM | DTPM | DTDM | SCHH | SCHT | SCHS | SCHD | SCHU | SCDH | SCDN | SCDP | SCDA | DTCE | DTCW | PEDS | PEDP | PEDE | DTRL | DTTP | DTTD |
|---|---|---|---|---|---|---|---|---|---|---|---|---|---|---|---|---|---|---|---|---|
| DTAM | 0.00000 | 0.50000 | 0.50000 | 0.33333 | 0.33333 | 0.33333 | 0.33333 | 0.33333 | 0.68542 | 0.33333 | 0.33333 | 0.33333 | 0.71429 | 0.28083 | 0.46154 | 0.33333 | 0.33333 | 0.33333 | 0.33333 | 0.33333 |
| DTPM | 0.50000 | 0.00000 | 0.50000 | 0.33333 | 0.33333 | 0.33333 | 0.33333 | 0.33333 | 0.23441 | 0.33333 | 0.33333 | 0.33333 | 0.14286 | 0.58415 | 0.46154 | 0.33333 | 0.33333 | 0.33333 | 0.33333 | 0.33333 |
| DTDM | 0.50000 | 0.50000 | 0.00000 | 0.33333 | 0.33333 | 0.33333 | 0.33333 | 0.33333 | 0.08017 | 0.33333 | 0.33333 | 0.33333 | 0.14286 | 0.13501 | 0.07692 | 0.33333 | 0.33333 | 0.33333 | 0.33333 | 0.33333 |
| SCHH | 0.00000 | 0.00000 | 0.00000 | 0.00000 | 0.00000 | 0.00000 | 0.00000 | 0.00000 | 0.00000 | 0.00000 | 0.00000 | 0.00000 | 0.00000 | 0.00000 | 0.00000 | 0.00000 | 0.00000 | 0.00000 | 0.00000 | 0.00000 |
| SCHT | 0.00000 | 0.00000 | 0.00000 | 0.00000 | 0.00000 | 0.00000 | 0.00000 | 0.00000 | 0.00000 | 0.00000 | 0.00000 | 0.00000 | 0.00000 | 0.00000 | 0.00000 | 0.00000 | 0.00000 | 0.00000 | 0.00000 | 0.00000 |
| SCHS | 0.00000 | 0.00000 | 0.00000 | 0.00000 | 0.00000 | 0.00000 | 0.00000 | 0.00000 | 0.00000 | 0.00000 | 0.00000 | 0.00000 | 0.00000 | 0.00000 | 0.00000 | 0.00000 | 0.00000 | 0.00000 | 0.00000 | 0.00000 |
| SCHD | 0.00000 | 0.00000 | 0.00000 | 0.00000 | 0.00000 | 0.00000 | 0.00000 | 0.00000 | 0.00000 | 0.00000 | 0.00000 | 0.00000 | 0.00000 | 0.00000 | 0.00000 | 0.00000 | 0.00000 | 0.00000 | 0.00000 | 0.00000 |
| SCHU | 0.00000 | 0.00000 | 0.00000 | 0.67259 | 0.64686 | 0.66282 | 0.65559 | 0.00000 | 0.66220 | 0.75000 | 0.75000 | 0.75000 | 0.66220 | 0.66220 | 0.66220 | 0.66220 | 0.66220 | 0.66220 | 0.66220 | 0.66046 |
| SCDH | 0.00000 | 0.00000 | 0.00000 | 0.07588 | 0.10047 | 0.08072 | 0.11069 | 0.15531 | 0.50000 | 0.00000 | 0.12500 | 0.12500 | 0.07843 | 0.07843 | 0.07843 | 0.07843 | 0.07843 | 0.07843 | 0.07843 | 0.07448 |
| SCDN | 0.00000 | 0.00000 | 0.00000 | 0.11969 | 0.09786 | 0.09124 | 0.09779 | 0.13592 | 0.25000 | 0.12500 | 0.00000 | 0.12500 | 0.12968 | 0.12968 | 0.12968 | 0.12968 | 0.12968 | 0.12968 | 0.12968 | 0.12831 |
| SCDP | 0.00000 | 0.00000 | 0.00000 | 0.13184 | 0.15481 | 0.13039 | 0.13592 | 0.09124 | 0.25000 | 0.12500 | 0.12500 | 0.00000 | 0.12968 | 0.12968 | 0.12968 | 0.12968 | 0.12968 | 0.12968 | 0.12968 | 0.13675 |
| SCDA | 0.25000 | 0.11111 | 0.14286 | 0.73850 | 0.83333 | 0.88889 | 0.14286 | 0.00000 | 0.00000 | 0.00000 | 0.00000 | 0.00000 | 1.00000 | 0.00000 | 0.00000 | 0.00000 | 0.00000 | 0.00000 | 0.00000 | 0.00000 |
| DTCE | 0.75000 | 0.88889 | 0.85714 | 0.26150 | 0.16667 | 0.11111 | 0.85714 | 0.00000 | 0.00000 | 0.00000 | 0.00000 | 0.00000 | 1.00000 | 0.00000 | 0.00000 | 0.00000 | 0.00000 | 1.00000 | 0.50000 | 0.50000 |
| DTCW | 0.00000 | 0.00000 | 0.00000 | 0.00000 | 0.00000 | 0.00000 | 0.00000 | 0.00000 | 0.00000 | 0.00000 | 0.00000 | 0.00000 | 0.00000 | 1.00000 | 0.00000 | 0.00000 | 0.00000 | 1.00000 | 0.50000 | 0.50000 |
| PEDS | 0.00000 | 0.00000 | 0.00000 | 0.00000 | 0.00000 | 0.00000 | 0.00000 | 0.00000 | 0.00000 | 0.00000 | 0.00000 | 0.00000 | 0.00000 | 0.00000 | 0.00000 | 0.50000 | 0.50000 | 0.00000 | 0.00000 | 0.00000 |
| PEDP | 0.00000 | 0.00000 | 0.00000 | 0.00000 | 0.00000 | 0.00000 | 0.00000 | 0.00000 | 0.00000 | 0.00000 | 0.00000 | 0.00000 | 0.00000 | 0.00000 | 0.50000 | 0.00000 | 0.50000 | 0.00000 | 0.00000 | 0.00000 |
| PEDE | 0.00000 | 0.00000 | 0.00000 | 0.00000 | 0.00000 | 0.00000 | 0.00000 | 0.00000 | 0.00000 | 0.00000 | 0.00000 | 0.00000 | 0.00000 | 0.00000 | 0.50000 | 0.50000 | 0.00000 | 0.00000 | 0.00000 | 0.00000 |
| DTRL | 1.00000 | 1.00000 | 1.00000 | 1.00000 | 1.00000 | 1.00000 | 1.00000 | 1.00000 | 0.50000 | 0.50000 | 0.56213 | 0.43787 | 0.50000 | 0.50000 | 0.50000 | 1.00000 | 0.50000 | 1.00000 | 0.00000 | 1.00000 |
| DTTP | 0.50000 | 0.50000 | 0.50000 | 0.58247 | 0.56868 | 0.50000 | 0.50000 | 0.50000 | 0.50000 | 0.50000 | 0.56213 | 0.50000 | 0.50000 | 0.50000 | 0.50000 | 0.50000 | 0.50000 | 0.50000 | 1.00000 | 0.10000 |
| DTTD | 0.50000 | 0.50000 | 0.50000 | 0.41753 | 0.43133 | 0.50000 | 0.50000 | 0.50000 | 0.50000 | 0.50000 | 0.43787 | 0.50000 | 0.50000 | 0.50000 | 0.50000 | 0.50000 | 0.50000 | 0.50000 | 0.00000 | 0.00000 |

### Weighted supermatrix

| Nodes | DTAM | DTPM | DTDM | SCHH | SCHT | SCHS | SCHD | SCHU | SCDH | SCDN | SCDP | SCDA | DTCE | DTCW | PEDS | PEDP | PEDE | DTRL | DTTP | DTTD |
|---|---|---|---|---|---|---|---|---|---|---|---|---|---|---|---|---|---|---|---|---|
| DTAM | 0.00000 | 0.17987 | 0.17987 | 0.07754 | 0.07754 | 0.07754 | 0.07754 | 0.07754 | 0.17205 | 0.08367 | 0.08367 | 0.08367 | 0.26617 | 0.10465 | 0.11464 | 0.08279 | 0.08279 | 0.09926 | 0.10634 | 0.10634 |
| DTPM | 0.17987 | 0.00000 | 0.17987 | 0.07754 | 0.07754 | 0.07754 | 0.07754 | 0.07754 | 0.05884 | 0.08367 | 0.08367 | 0.08367 | 0.05323 | 0.21768 | 0.10464 | 0.08279 | 0.08279 | 0.09926 | 0.10634 | 0.10634 |
| DTDM | 0.17987 | 0.17987 | 0.00000 | 0.07754 | 0.07754 | 0.07754 | 0.07754 | 0.07754 | 0.02012 | 0.08367 | 0.08367 | 0.08367 | 0.05323 | 0.05031 | 0.01911 | 0.08279 | 0.08279 | 0.09926 | 0.10634 | 0.10634 |
| SCHH | 0.00000 | 0.00000 | 0.00000 | 0.00000 | 0.00000 | 0.00000 | 0.00000 | 0.00000 | 0.00000 | 0.00000 | 0.00000 | 0.00000 | 0.00000 | 0.00000 | 0.00000 | 0.00000 | 0.00000 | 0.00000 | 0.00000 | 0.00000 |
| SCHT | 0.00000 | 0.00000 | 0.00000 | 0.00000 | 0.00000 | 0.00000 | 0.00000 | 0.00000 | 0.00000 | 0.00000 | 0.00000 | 0.00000 | 0.00000 | 0.00000 | 0.00000 | 0.00000 | 0.00000 | 0.00000 | 0.00000 | 0.00000 |
| SCHS | 0.00000 | 0.00000 | 0.00000 | 0.00000 | 0.00000 | 0.00000 | 0.00000 | 0.00000 | 0.00000 | 0.00000 | 0.00000 | 0.00000 | 0.00000 | 0.00000 | 0.00000 | 0.00000 | 0.00000 | 0.00000 | 0.00000 | 0.00000 |
| SCHD | 0.00000 | 0.00000 | 0.00000 | 0.00000 | 0.00000 | 0.00000 | 0.00000 | 0.00000 | 0.00000 | 0.00000 | 0.00000 | 0.00000 | 0.00000 | 0.00000 | 0.00000 | 0.00000 | 0.00000 | 0.00000 | 0.00000 | 0.00000 |
| SCHU | 0.00000 | 0.00000 | 0.00000 | 0.04446 | 0.04276 | 0.04382 | 0.04334 | 0.00000 | 0.05565 | 0.10022 | 0.10371 | 0.10371 | 0.10371 | 0.10371 | 0.10371 | 0.10371 | 0.10371 | 0.17058 | 0.12259 | 0.12227 |
| SCDH | 0.00000 | 0.00000 | 0.00000 | 0.00502 | 0.00664 | 0.00534 | 0.00732 | 0.01027 | 0.03710 | 0.00000 | 0.00928 | 0.01187 | 0.01228 | 0.01228 | 0.01187 | 0.01027 | 0.01027 | 0.01228 | 0.01452 | 0.01379 |
| SCDN | 0.00000 | 0.00000 | 0.00000 | 0.00732 | 0.00534 | 0.00664 | 0.00502 | 0.01027 | 0.01027 | 0.00928 | 0.00000 | 0.01187 | 0.01228 | 0.01228 | 0.01187 | 0.01027 | 0.01027 | 0.01228 | 0.01452 | 0.01379 |

|  | | | | | | | | | | | | | | | | |
|---|---|---|---|---|---|---|---|---|---|---|---|---|---|---|---|---|
| SCDP | 0.00000 | 0.00000 | 0.00000 | 0.00791 | 0.00647 | 0.00833 | 0.01855 | 0.00928 | 0.00000 | 0.00928 | 0.00000 | 0.01963 | 0.02031 | 0.03341 | 0.02401 | 0.02375 |
| SCDA | 0.00000 | 0.00000 | 0.00000 | 0.00872 | 0.01023 | 0.00898 | 0.01855 | 0.00603 | 0.00928 | 0.00603 | 0.00000 | 0.01963 | 0.02031 | 0.03341 | 0.02401 | 0.02532 |
| DTCE | 0.03867 | 0.01719 | 0.02210 | 0.14283 | 0.12658 | 0.08570 | 0.08570 | 0.08548 | 0.08548 | 0.08548 | 0.08548 | 0.00040 | 0.05586 | 0.05586 | 0.12247 | 0.07603 | 0.07603 |
| DTCW | 0.11601 | 0.13749 | 0.13258 | 0.04482 | 0.02857 | 0.08570 | 0.08570 | 0.08548 | 0.08548 | 0.08548 | 0.08548 | 0.00040 | 0.05586 | 0.05586 | 0.12247 | 0.07603 | 0.07603 |
| PEDS | 0.00000 | 0.00000 | 0.00000 | 0.00000 | 0.00000 | 0.00000 | 0.00000 | 0.00000 | 0.00000 | 0.00000 | 0.10973 | 0.00000 | 0.10973 | 0.05586 | 0.05586 | 0.10973 | 0.00000 | 0.00000 |
| PEDP | 0.00000 | 0.00000 | 0.00000 | 0.00000 | 0.00000 | 0.00000 | 0.00000 | 0.00000 | 0.00000 | 0.00000 | 0.10973 | 0.00000 | 0.10973 | 0.05586 | 0.05586 | 0.00000 | 0.00000 |
| PEDE | 0.00000 | 0.00000 | 0.00000 | 0.00000 | 0.00000 | 0.00000 | 0.00000 | 0.00000 | 0.00000 | 0.00000 | 0.10973 | 0.10973 | 0.00000 | 0.05586 | 0.05586 | 0.00000 | 0.00000 |
| DTRL | 0.30446 | 0.30446 | 0.30446 | 0.32196 | 0.32196 | 0.32196 | 0.30905 | 0.30905 | 0.30905 | 0.15212 | 0.15212 | 0.00040 | 0.00000 | 0.00000 | 0.13719 | 0.13719 |
| DTTP | 0.09056 | 0.09056 | 0.09056 | 0.12111 | 0.10396 | 0.11824 | 0.10396 | 0.09739 | 0.09739 | 0.10949 | 0.09739 | 0.09739 | 0.23761 | 0.23761 | 0.05586 | 0.05586 | 0.09985 | 0.09985 |
| DTTD | 0.09056 | 0.09056 | 0.09056 | 0.10396 | 0.08681 | 0.08968 | 0.10396 | 0.08529 | 0.09739 | 0.09739 | 0.09739 | 0.23761 | 0.23761 | 0.05586 | 0.05586 | 0.20661 | 0.00000 |

*Limiting supermatrix*

| Nodes | DTAM | DTPM | DTDM | SCHH | SCHT | SCHS | SCHD | SCHU | SCDH | SCDN | SCDP | SCDA | DTCE | DTCW | PEDS | PEDP | PEDE | DTRL | DTTP | DTTD |
|---|---|---|---|---|---|---|---|---|---|---|---|---|---|---|---|---|---|---|---|---|
| DTAM | 0.12059 | 0.12059 | 0.12059 | 0.12059 | 0.12059 | 0.12059 | 0.12059 | 0.12059 | 0.12059 | 0.12059 | 0.12059 | 0.12059 | 0.12059 | 0.12059 | 0.12059 | 0.12059 | 0.12059 | 0.12059 | 0.12059 | 0.12059 |
| DTPM | 0.11174 | 0.11174 | 0.11174 | 0.11174 | 0.11174 | 0.11174 | 0.11174 | 0.11174 | 0.11174 | 0.11174 | 0.11174 | 0.11174 | 0.11174 | 0.11174 | 0.11174 | 0.11174 | 0.11174 | 0.11174 | 0.11174 | 0.11174 |
| DTDM | 0.09638 | 0.09638 | 0.09638 | 0.09638 | 0.09638 | 0.09638 | 0.09638 | 0.09638 | 0.09638 | 0.09638 | 0.09638 | 0.09638 | 0.09638 | 0.09638 | 0.09638 | 0.09638 | 0.09638 | 0.09638 | 0.09638 | 0.09638 |
| SCHH | 0.00000 | 0.00000 | 0.00000 | 0.00000 | 0.00000 | 0.00000 | 0.00000 | 0.00000 | 0.00000 | 0.00000 | 0.00000 | 0.00000 | 0.00000 | 0.00000 | 0.00000 | 0.00000 | 0.00000 | 0.00000 | 0.00000 | 0.00000 |
| SCHT | 0.00000 | 0.00000 | 0.00000 | 0.00000 | 0.00000 | 0.00000 | 0.00000 | 0.00000 | 0.00000 | 0.00000 | 0.00000 | 0.00000 | 0.00000 | 0.00000 | 0.00000 | 0.00000 | 0.00000 | 0.00000 | 0.00000 | 0.00000 |
| SCHS | 0.00000 | 0.00000 | 0.00000 | 0.00000 | 0.00000 | 0.00000 | 0.00000 | 0.00000 | 0.00000 | 0.00000 | 0.00000 | 0.00000 | 0.00000 | 0.00000 | 0.00000 | 0.00000 | 0.00000 | 0.00000 | 0.00000 | 0.00000 |
| SCHD | 0.00000 | 0.00000 | 0.00000 | 0.00000 | 0.00000 | 0.00000 | 0.00000 | 0.00000 | 0.00000 | 0.00000 | 0.00000 | 0.00000 | 0.00000 | 0.00000 | 0.00000 | 0.00000 | 0.00000 | 0.00000 | 0.00000 | 0.00000 |
| SCHU | 0.00000 | 0.00000 | 0.00000 | 0.00000 | 0.00000 | 0.00000 | 0.00000 | 0.00000 | 0.00000 | 0.00000 | 0.00000 | 0.00000 | 0.00000 | 0.00000 | 0.00000 | 0.00000 | 0.00000 | 0.00000 | 0.00000 | 0.00000 |
| SCDH | 0.07495 | 0.07495 | 0.07495 | 0.07495 | 0.07495 | 0.07495 | 0.07495 | 0.07495 | 0.07495 | 0.07495 | 0.07495 | 0.07495 | 0.07495 | 0.07495 | 0.07495 | 0.07495 | 0.07495 | 0.07495 | 0.07495 | 0.07495 |
| SCDN | 0.01159 | 0.01159 | 0.01159 | 0.01159 | 0.01159 | 0.01159 | 0.01159 | 0.01159 | 0.01159 | 0.01159 | 0.01159 | 0.01159 | 0.01159 | 0.01159 | 0.01159 | 0.01159 | 0.01159 | 0.01159 | 0.01159 | 0.01159 |
| SCDP | 0.01583 | 0.01583 | 0.01583 | 0.01583 | 0.01583 | 0.01583 | 0.01583 | 0.01583 | 0.01583 | 0.01583 | 0.01583 | 0.01583 | 0.01583 | 0.01583 | 0.01583 | 0.01583 | 0.01583 | 0.01583 | 0.01583 | 0.01583 |
| SCDA | 0.01601 | 0.01601 | 0.01601 | 0.01601 | 0.01601 | 0.01601 | 0.01601 | 0.01601 | 0.01601 | 0.01601 | 0.01601 | 0.01601 | 0.01601 | 0.01601 | 0.01601 | 0.01601 | 0.01601 | 0.01601 | 0.01601 | 0.01601 |
| DTCE | 0.05748 | 0.05748 | 0.05748 | 0.05748 | 0.05748 | 0.05748 | 0.05748 | 0.05748 | 0.05748 | 0.05748 | 0.05748 | 0.05748 | 0.05748 | 0.05748 | 0.05748 | 0.05748 | 0.05748 | 0.05748 | 0.05748 | 0.05748 |
| DTCW | 0.09088 | 0.09088 | 0.09088 | 0.09088 | 0.09088 | 0.09088 | 0.09088 | 0.09088 | 0.09088 | 0.09088 | 0.09088 | 0.09088 | 0.09088 | 0.09088 | 0.09088 | 0.09088 | 0.09088 | 0.09088 | 0.09088 | 0.09088 |
| PEDS | 0.00000 | 0.00000 | 0.00000 | 0.00000 | 0.00000 | 0.00000 | 0.00000 | 0.00000 | 0.00000 | 0.00000 | 0.00000 | 0.00000 | 0.00000 | 0.00000 | 0.00000 | 0.00000 | 0.00000 | 0.00000 | 0.00000 | 0.00000 |
| PEDP | 0.00000 | 0.00000 | 0.00000 | 0.00000 | 0.00000 | 0.00000 | 0.00000 | 0.00000 | 0.00000 | 0.00000 | 0.00000 | 0.00000 | 0.00000 | 0.00000 | 0.00000 | 0.00000 | 0.00000 | 0.00000 | 0.00000 | 0.00000 |
| PEDE | 0.00000 | 0.00000 | 0.00000 | 0.00000 | 0.00000 | 0.00000 | 0.00000 | 0.00000 | 0.00000 | 0.00000 | 0.00000 | 0.00000 | 0.00000 | 0.00000 | 0.00000 | 0.00000 | 0.00000 | 0.00000 | 0.00000 | 0.00000 |
| DTRL | 0.16904 | 0.16904 | 0.16904 | 0.16904 | 0.16904 | 0.16904 | 0.16904 | 0.16904 | 0.16904 | 0.16904 | 0.16904 | 0.16904 | 0.16904 | 0.16904 | 0.16904 | 0.16904 | 0.16904 | 0.16904 | 0.16904 | 0.16904 |
| DTTP | 0.11792 | 0.11792 | 0.11792 | 0.11792 | 0.11792 | 0.11792 | 0.11792 | 0.11792 | 0.11792 | 0.11792 | 0.11792 | 0.11792 | 0.11792 | 0.11792 | 0.11792 | 0.11792 | 0.11792 | 0.11792 | 0.11792 | 0.11792 |
| DTTD | 0.11760 | 0.11760 | 0.11760 | 0.11760 | 0.11760 | 0.11760 | 0.11760 | 0.11760 | 0.11760 | 0.11760 | 0.11760 | 0.11760 | 0.11760 | 0.11760 | 0.11760 | 0.11760 | 0.11760 | 0.11760 | 0.11760 | 0.11760 |

limiting supermatrix in Table 3.17, $w_{C_{Plan},i} = (0.120594, 0.111735, 0.096383)$, so $W_i = (0.366867, 0.339917, 0.293216)$; as a result, the best demolition plan is DTAM.

### 3.4.5 Comparison between DEMAP and DEMAN

Both DEMAP and DEMAN provided the same conclusion that the demolition plan using DTAM is the best demolition plan. Besides the selected demolition plan, it is also noticed that priority queues of these three demolition plan alternatives are also equivalent (refer to Table 3.18).

The comparison result implies that the most preferable demolition plan regulates the demolition practice with the least requirement on machinery, and the lowest risk ratios of health and safety for people on and off site, because of the attributes of demolition plan alternatives listed in Table 3.14. This result also indicates both DEMAP and DEMAN can provide a quite reasonable comparison result for environmental-conscious demolition.

Although the DEMAN appears to provide a more precise result than the DEMAP due to its load of performing pairwise comparisons between clusters and between nodes, the difference between priority weights of DTAM and DTPM in the DEMAN is not as absolutely clear as those in the DEMAP. There are two possible explanations for this result. One explanation is that there is a risk of getting results which provide unrealistic rankings when ANP is applied comparing with the results from AHP (Salomon and Montevechi 2001). On the contrary, another explanation is that the difference of advantages between DTAM and DTPM is not significant indeed. For example, there is difference between DTAM and DTPM in three attributes: the SCDH, DTCE, and DTCW (refer to Table 3.14). Because the DTAM is preferable to DTPM in SCDH and DTCE, and is inferior to DTPM in DTCW, there is no absolute advantage in DTAM; and the authors prefer to agree to the second explanation. However, in order to prove that the DEMAN can provide a more precise result than the DEMAP, the authors suggest further case studies other than the demonstration project used in this study.

Moreover, according to the calculation results of priority weight (refer to Table 3.18), it is also noticed that there is no demolition plan with a priority weight over 0.5 in the two MCDM models. There are also two possible reasons.

*Table 3.18* A comparison between two MCDM models using priority weight

| MCDM model | No. of nodes | Synthesized priority weight | | | Selection |
|---|---|---|---|---|---|
| | | DTAM plan | DTPM plan | DTDM plan | |
| DEMAP | 17 | 0.490 | 0.318 | 0.192 | DTAM plan |
| DEMAN | 20 | 0.367 | 0.340 | 0.293 | DTAM plan |

One possible reason is that none of these three demolition plans has significant advantage over others in this demonstration project, whilst another possible reason is that the evaluation criteria used in the DEMAP and the DEMAN cannot significantly distinguish these three demolition techniques by using the attributes defined. As a result, further researches will focus on the evaluation of the two MCDM models in different demolition projects and modification of the evaluation criteria used in the two MCDM models.

### 3.4.6 Summary

There are two contributions in this section. The first one is that the authors successfully transplant the intact evaluation criteria of selecting the best demolition technique into both DEMAP model and DEMAN model, and another one is that the authors make a comparison between the two MCDM models by using a case from a demonstration project. Although the evaluation criteria of selecting the best demolition technique can be transplanted in the DEMAP model and the DEMAN model, and both of these two MCDM models can work well for selecting the best demolition plan, there are also some problems. For example, no priority weight of a demolition plan is over 0.5 according to the calculation results, and the differences of priority weight among plan alternatives are small, especially for the DEMAN model. The authors also discussed the possible reasons related to these problems and the direction of further research.

## 3.5 Conclusions and discussions

A quantitative approach to construction pollution management by introducing parameters of CPI and pollution and hazard magnitude $h_i$ has been proposed. By using these parameters, a method to predict the distribution of accumulated pollution level generated from construction operations is presented. It is suggested that if the pollution level exceeds the allowable limit, then construction activities need to be re-scheduled to "spread" the polluting emissions. In doing so, polluting emission is treated as a pseudo resource, and then applied to a GA-based levelling technique to re-schedule the project activities. The GA allows the user to concurrently minimize fluctuations and period of resource use by assigning different priorities to project activities. Experimental results indicate that GA-enhanced resource levelling performs better than the traditional resource levelling method used in Microsoft Project©.

As a matter of fact, the proposed method for controlling construction pollution is an effective tool that can be used by project managers to reduce the level of pollution generated from a project at a certain period of time. This method is useful when there is no other way to reduce the level of pollution. However, it is necessary to point out that the method proposed here can only redistribute the amount of pollution over project duration so that at any specific period of time, the level of pollution will not exceed the legal limit. In order to reduce the overall amount of pollution, other methods, such as alternative construction technologies and new materials, have to be applied.

This chapter also presents an env.Plan method for environmental-conscious construction planning when plan alternatives need to be selected for reducing adverse environmental impacts in construction, especially after CPI levelling. The env.Plan method was constructed and illustrated using ANP, and both simplified env.Plan model and complicated env.Plan model are developed. The simplified model consists of 4 clusters and 15 corresponding nodes, while the complicated model consists of 4 clusters and 35 corresponding nodes. In addition, performances of the two models are compared and the results indicated that while the complicated model yielded accurate results, the simplified model is easy to use.

The env.Plan method is put forward based on an ANP model which contains feedback and self-loops among the clusters (refer to Figure 3.10), but no control structure. However, there is an implicit control criterion with respect to which all judgements are made in the env.Plan model: environmental impact. The supermatrix computations are conducted for the overall priorities of plan alternatives, which are obtained by synthesizing the priorities of the alternatives from all the subnetworks of the ANP model. Finally, the synthesized priority weight $W_i$ is used to distinguish the degree of potential environmental impacts due to the implementation of a construction plan.

However, problems also exist in the env.Plan method; for example, the reliability of the three clusters – EA Factors ($C_2$), EU Factors ($C_3$) and EF Factors ($C_4$) – and their nodes cannot be measured. As the sorting criteria rely on the calculation results of the $EI_i$, subjective judgements can influence the accuracy of the system. Further studies are therefore needed to investigate these issues.

The ANP is employed here to realize the purpose of demolition plan selection. It is concluded that the ANP is a viable and capable tool for selecting the best demolition plan by using the same set of evaluation criteria transplanted from the AHP model developed in previous research. However, although the ANP has the ability to measure relationships among selection criteria and their subcriteria, which is normally ignored in the AHP, the authors also conclude that it should be examined if the ANP model can provide a more accurate result in further research.

# Chapter 4

# Effective control at construction stage

## 4.1 Introduction

Construction waste is a serious environmental problem in many large cities. According to statistical data, C&D debris frequently makes up 10–30 percent of the waste received at many landfill sites around the world (Fishbein 1998). However, in Hong Kong, an average of 7,030 tons of C&D waste were disposed of at landfills everyday in 1998, representing about 42% of total waste intake at landfills, and most of which can be reclaimed; and in 1999, there were 7890 tons of C&D waste disposed of at landfills every day, representing about 44% of total waste intake at landfills (HKEPD 1999a,b,c,d, 2000a,b,c,d). In contrast to the percentage in other advanced countries, for example, C&D debris makes up only 12% of the waste received at Metro Park East Sanitary Landfill of Iowa State in the United States (MWA 2000); the quantity of C&D waste in Hong Kong is much higher. As there are increasing demands on residential buildings in Hong Kong, a 13-year production program had been established by the Hong Kong SAR government in 1998, which has been rolled forward to produce an average of 50,000 flats in the public sector and 35,000 flats in the private sector each year (HB 2000). So how to reduce construction waste is becoming more important in Hong Kong.

There have been many research efforts for construction waste control in Hong Kong. For example, a study that investigated construction waste generated from public housing projects in Hong Kong was conducted in 1992 (Cheung *et al.* 1993). Methods for construction waste minimization in Hong Kong were also discussed by (Poon *et al.* 1996). These waste minimization methods emphasize the use of modern technologies in building construction, such as precast concrete, steel form and scaffold, drywall partition panel, etc. However, surveys show that local construction firms in Hong Kong feel it is expensive to use new machinery and automation (Ho 1997); most (68–85%) local construction firms agree to adopt low-waste techniques only when they are demanded by the designers, the specifications, or the clients (Poon and Ng 1999). As a result, construction waste control in Hong Kong is still a major problem to be solved.

Previous practice and studies have established a set of waste prevention strategies considered in building construction. These strategies mainly involve the effective coordination of materials management, including efficient purchase and ordering of materials; efficient timing and delivery; efficient storage; and the use of materials to minimize loss, maximize re-use, prevent undoing and redoing, and reduce packaging waste, etc. (Fishbein 1998). The management of on-site waste is thus emphasized on executing a waste management plan for each construction and demolition site (Coventry *et al.* 1999). As another important factor, design coordination has a major impact on waste generation. Incorrect or unconstructable designs result in significant amounts of wastes. A study on the relationship between causes and costs of rework indicates that, among other factors, design coordination is predominantly important (Love and Li 2000). However, as the housing projects in Hong Kong adopt a series of standard designs developed by the Housing Authority of the Hong Kong SAR, the effect of design coordination is minimized, if not negligible. Therefore, in this study, the impact of design coordination on waste generation is not considered.

The objective of this chapter is to present an on-site material management scheme using an incentive reward program (IRP) to control and reduce construction wastes. The scheme is designed to encourage construction workers, who are directly involved in producing construction wastes, to reduce wastes by rewarding them based on the amounts and values of the materials they saved. The bar-coding technique is used to facilitate easy data recording and transfer.

## 4.2 Generation of construction wastes

Although there is no generally accepted definition, construction waste can be loosely defined as the debris of C&D (U.S.EPA 2000). Specifically, construction waste refers to solid waste containing no liquids and hazardous substances, largely inert waste, resulting from the process of construction of structures, including buildings of all types (both residential and nonresidential) as well as roads and bridges. Construction waste does not include clean-up materials contaminated with hazardous substances, friable asbestos-containing materials, lead, waste paints, solvents, sealers, adhesives, living garbage, furniture, appliances, or similar materials.

Although it is difficult to give exact figures of construction wastes generated on a construction site, it is estimated that as much as 10–30% construction materials are wasted (Stone 1983; Fishbein 1998). Data obtained from specialty contractors in USA, UK, mainland China, Hong Kong, Brazil, and Korea present a comparison of construction wastes generated from construction industries in these countries/regions, which is displayed in Table 4.1.

The authors conducted a construction waste survey, in which an on-the-spot investigation about construction waste generation in residential projects in Hong Kong is planned, and we aim to put forward a reasonable scheme to solve the problem of construction waste generation. In our construction site study, both major contractors and clients are selected on account of their technologies and projects that are widely representative in the Hong Kong construction industry. The contractors are Yau Lee Construction Co., Ltd and Hung Hom Construction Co., Ltd; and the clients are the Hong Kong Housing Authority and Sun Hung Kai

*Table 4.1* Average on-site wastage rate of construction materials

| Material | Average wastage (%) | | | | | |
|---|---|---|---|---|---|---|
| | USA | UK | Mainland China | Hong Kong | Brazil | Seoul |
| Brick/Block | 3.5 | 4.5 | 2.0 | N/A | 17.5 | 3.0 |
| Concrete | 7.5 | 2.5 | 2.5 | 6.7 | 7.0 | 1.5 |
| Drywall | 7.5 | 5.0 | N/S | 9.0 | N/S | N/S |
| Formwork | 10.0 | N/S | 7.5 | 4.6 | N/S | 16.7 |
| Glass | N/S | N/S | 0.8 | 2.3 | N/S | 6.0 |
| Mortar | 3.5 | N/S | 5.0 | 3.2 | 46.0 | 0.3 |
| Nail | 5.0 | N/S | N/S | N/A | N/S | N/S |
| Rebar | 5.0 | N/S | 3.0 | 8.0 | 21.0 | N/S |
| Tile | 6.5 | 5.0 | N/S | 6.3 | 8.0 | 2.5 |
| Wallpaper | 10.0 | N/S | N/S | N/A | N/S | 11.0 |
| Wood | 16.5 | 6.0 | N/S | 45.0 | 32.0 | 13.0 |

Notes
1 N/S = Not specified, N/A = Not available;
2 Reference: USA (Schuette and Liska 1994), UK (Skoyles 1992; Frics 1996); Mainland China (Zhu 1996), Hong Kong (Site surveys), Brazil (Bossink and Brouwers 1996), and Seoul (Seo and Hwang 1999).

Properties Co., Ltd. Two representative public housing projects and one private housing project are selected for the survey. Of the two public housing projects, one is a public housing project (Phase 4) on Po Lam Road, Kowloon, and another is a public housing project (Phase 1) on Cheung She Wan West, New Territory; and the private housing project is Royal Peninsula adjacent to the KCR Kowloon Terminus, Kowloon. The construction sites study was conducted during the stage of superstructure works until finish works, from November 1999 to April 2000.

The typical public housing block in Hong Kong is a multi-floor reinforced concrete (RC) residential building with about 40 floors. The construction technologies of public housing block buildings are summarized in Table 4.2.

According to our site surveys of superstructure works of the residential projects, construction waste generated mainly includes wastage of cement, concrete rubbles, drywall scraps, wood scraps, rebar scraps, concrete block scraps, plastic conduit tailings, material packing and containers, nails, and other unused materials. For example, a site survey of public housing projects shows (refer to Table 4.3) that different construction processes can generate different construction waste, and it is similar in private residential project.

The reason why different construction wastes are generated from different construction processes can be divided into four sections, including construction technology, management, method, materials, and workers.

### 4.2.1 Construction technology

Both prefabrication technology and in situ technology of reinforced concrete are used in residential projects. The prefabrication technology generates almost no

Table 4.2 Construction technologies of public housing block in HK

| Stage | Technologies |
|---|---|
| Site formation and clearance works | Demolition, site levelling |
| Foundation works | Precast RC pile, excavation, in situ RC foundation |
| Superstructure works | Precast RC external wall panel, in situ RC load-bearing wall, corridor and slab, semi-precast RC slab, precast concrete internal drywall, precast RC staircase, precast concrete block |
| Finish works | In situ external and internal plastering and coating, external wall and floor tiling |
| Other works | Batching plant, tyrewasher system, precast plant, transportation |

Table 4.3 Construction waste generated from construction processes

| Construction process | Construction waste | | | | | | | | |
|---|---|---|---|---|---|---|---|---|---|
| | Concrete rubble | Drywall scrap | Block scrap | Cement wastage | Wood scrap | Rebar tailing | Nail | Plastic conduit tailing | Material packing and container |
| Fix wall rebar | | | | | | ✓ | | ✓ | |
| Place precast | | | | | | | | | ✓ |
| Place wall form | | | | ✓ | ✓* | | | | |
| Concrete wall | ✓ | | | ✓ | | | | | ✓ |
| Strip wall form | ✓ | | | | | | | | |
| Place precast | | | | | | | | | |
| Fix timber slab | | | ✓ | ✓ | | | ✓ | | |
| Fix slab rebar | | | | | | ✓ | | ✓ | |
| Concrete slab | ✓ | | | ✓ | | | | | ✓ |
| Fix drywall | | ✓ | | | | | | | |
| Bond block | | | ✓ | | | | | | |

Note
*When through-wall sleeve cannot be fixed easily, wall rebar will be cut.

construction waste because there is no need to use rebar, wood form, and in situ concrete, etc. on the site. On the other hand, in situ technology generates wastage of rebar, timber, and concrete, etc. during the process of construction, which is difficult to prevent on site.

### 4.2.2 Management method

In the site survey, it has been noticed that most construction wastes were generated due to the disorder of construction site layout. In some sites, materials and tools were placed everywhere, and as a result some unused materials and tools were messed up with the wastes and were eventually removed as wastes. Therefore, methods for managing and controlling wastes influence the amounts of wastes generated on site. For example, the introduction of waste storage containers (refer to Table 4.4) help to sort out various types of wastes. These sorted wastes are easy to recycle and re-use.

Obviously, these waste management methods can systematically sort out construction wastes on the sites; they cannot reduce construction wastes generated from every process. For example, the drywall board is a kind of solid slab, when workers fix pipelines, they cut the slab as they like and do not think about the amount of cuts and concrete fillings, and waste is thus generated. In the current management practice, the site waste manager's duty is only to collect the wastes and ensure the site is neat. In order to reduce the wastes, it is necessary to make innovations in the management of materials and equipments such as training to workers to reduce avoidable wastes, and due reward to workers for the good practices in cutting down wastages. One reason why the current management method cannot effectively reduce waste on construction sites is that it cannot effectively control the generation of construction waste due to the faults of construction techniques, building materials, workers, etc. From this point of view, innovative management methods are required to decrease any fault in waste reduction.

### 4.2.3 Materials

Two kinds of construction wastes originated from construction materials: materials packaging and materials wastage discarded on the construction site. Because construction packaging made of kraft paper and timber, and cartons are

Table 4.4 Current measures for construction waste management on site

| Construction waste | Management measure |
| --- | --- |
| Rebar | Useless rebar collection skip |
| Concrete, Drywall, Block, Timber | Useless concrete transport pipe and collection skip |
| Water | On-site waste water treatment system |
| Other solid waste | On-site waste barrel |

necessary for packing construction materials such as cement, wall tile, mosaic, and concrete nail, etc., the packaging unavoidably becomes part of the waste when materials are unpacked on site.

### 4.2.4 Workers

Workers take part in construction activities, and the survey shows that their attitude towards construction operations can make a big difference in terms of construction waste generation. Specifically, it is observed that if workers do not handle the materials with sufficient care then they will waste more materials, and vice versa. It has been observed that one of the main causes of material waste generation is incorrect or careless use of materials by workers on site. These kinds of wastes can be avoided or reduced if workers are motivated to be more conscious and responsible.

## 4.3   Avoidable material wastes caused by workers

Without careful control and rewarding systems, construction workers may become careless in handling construction materials. As a result, reusable reinforcement bars, discarded half-bags of cement, discarded nails and timber pieces are often thrown around the sites. Table 4.5 gives examples of avoidable wastes caused by workers in public housing projects in Hong Kong.

Table 4.5 indicates that skill, enthusiasm, and collectivism are the main factors affecting the amounts of wastes produced by workers. Among these three factors, workers' attitude towards their work, including their enthusiasm and collectivism, is regarded as the most important aspect in terms of waste generation, while their skill levels are relatively less important. In other words, if workers do not take

Table 4.5 Avoidable wastes caused by workers in public housing projects in HK

| Construction process | Avoidable wastes caused by workers |
| --- | --- |
| Fix wall rebar | Extra processed rebar, arbitrarily cut rebar, abandoned rebar tailing, etc. |
| Place precast facade | Damaged facade board, broken scraps during erection |
| Place wall form | Arbitrarily cut and drilled plywood board, abandoned plywood board |
| Concrete wall | Left-over mixed concrete, excess concreting, etc. |
| Strip wall form | Damaged forms |
| Place precast slab | Damaged slab boards, broken scraps during erection |
| Fix timber slab | Arbitrarily cut and drilled plywood boards, abandoned plywood boards |
| Fix slab rebar | Extra processed rebar, arbitrarily cut rebar, abandoned rebar tailing, etc. |
| Concrete slab | Left-over mixed concrete, excessive concreting, etc. |
| Fix drywall | Arbitrarily cut drywall board, damaged drywall board, broken scraps, etc. |
| Bond block | Extra mortar, extra delivered blocks, cut and abandoned blocks, etc. |

care of what they are doing then more materials will be wasted. So it is important to establish an on-site construction material management system to encourage construction workers to use materials carefully, and to enhance their enthusiasm and collectivism by rewarding them based on their good performances in saving materials through reducing operational mistakes, returning unused materials for re-use or recycle, etc.

It has been pointed out that because most residential buildings adopt standard designs prepared by the Housing Authority of the Hong Kong SAR government, such as the Harmony series, and are constructed by similar methods such as 4-day cycle and 6-day cycle, factors such as design coordination do not have major impacts on the generation of material wastes. How to enhance workers' enthusiasm and collectivism in minimizing construction wastes thus becomes more important in residential projects in Hong Kong.

## 4.4 Incentive reward program

It was observed in our site surveys that construction materials are taken from the storage areas on site without effective control, and placed with poor organization, especially in large projects or during urgent construction processes. The construction-material control system to be established aims to provide an effective tool for the project manager to manage on-site materials, and to motivate workers to reduce material waste to its minimum.

Research on the relationship between motivation and productivity in the construction industry has been conducted over the last 40 years (Olomolaiye *et al.* 1998). Productivity is dependent upon motivation, and motivation is in turn dependent on productivity (Warren 1989). A comparison of labour productivity for masonry activities from seven countries, including Australia, Canada, England, Finland, Scotland, Sweden, and the United States, reveals that there is little difference in productivity in the seven countries despite significant differences in labour practices, and the principal difference is management influence (Thomas *et al.* 1992). This viewpoint is replenished with a case study focusing on the impact of material management on productivity, which shows that ineffective material management could incur losses of productivity (Thomas *et al.* 1990). On the other hand, a series of comparative evaluations of labour productivity rates amongst French, German, and UK construction contractors indicate that German workers are likely to be highly motivated (because they are highly paid and regarded to be on a par with people doing intellectual and scientific work), and hence, more productive (Proverbs *et al.* 1998). All these research results reinforce that higher motivation brings higher productivity.

According to Maslow's motivation theory (Warren 1989), beyond their safety and health needs, workers require both emotional and financial rewards for exercising self-discipline in handling construction materials. There are many forms of rewards and punishments for workers' performance measure (Nelson 1994). Among these positive and negative rewarding (punishing) methods, some have been used on construction sites. For example, the use of special motivational programs and financial incentive programs (FIPs) have been reported (Laufer and Jenkins 1982; Liska and Snell 1993; Carberry 1996; Olomolaiye *et al.* 1998). The FIP is an important method for motivating workers, and it has been proved to be effective in improving quality and reducing project time and cost (Laufer and Jenkins 1982). Furthermore, the FIP has been widely accepted as a performance-dependent monetary reward system in the construction industry (Merchant 1997). So the IRP developed in this study is designed based on the principle of FIP, in order to meet the demand of on-site construction material management.

Fairness is an important consideration in designing the IRP; less fairness or unfairness would result in the failure of the IRP and may even have adverse effects on a construction project. Before the IRP is implemented, its fairness should be examined carefully. There are two aspects of fairness in the IRP: one is fairness to workers, another is its fairness to the firm. Fairness to the firm is easy to investigate. Because the IRP relates to the amount of construction materials consumed on site, if the overall amounts of construction wastes are reduced, then the firm will benefit. So the firm should share the benefits (saved money) with the contributors – workers.

The fairness of the IRP to workers is different. Workers are normally organized into gangs or groups according to their trades or types of work. Material is normally shared within the group. If an amount of material waste is detected, who should be punished, or, if there is a reduction of waste, who should be rewarded – the person who is responsible for shifting material from storage, or the leader of the group? Based on discussions with the project managers and workers involved in the projects we surveyed, we decided to adopt a group-based IRP. In the group-based IRP, members of the group will be rewarded or punished equally should there be any reduction and increase of material wastes. Group-based rewards provide a common goal for group members and encourage cooperation among members to achieve a higher performance, and it avoids the difficulty in determining an individual's contribution (Laufer and Jenkins 1982; Merchant 1997).

In the group-based IRP, each working group has a group leader who is responsible for collecting all the materials needed by his group from the store keeper. The store keeper records the amount of materials taken by each group. When a group finishes its work, the group leader is also responsible for arranging any unused materials to be returned back to the store keeper for updating the records.

Once a construction operation is completed, the project manager can measure the amount of material waste reduced or increased by comparing the actual amount of material used by the group with the estimated amount. The actual amount of material used is recorded by the store keeper, while the estimated amount of material is prepared by the contractor's quantity surveyors. The estimated amount includes a percentage which is considered as a normal amount of waste on site. The percentage is determined based on the contractor's experience from the levels of wastes in past projects.

For a particular type of material $i$, the performance of group $j$ in terms of material wastage can be measured using Equation 4.1.

$$\Delta Q^i(j) = Q^i_{\text{estimated}}(j) - (Q^i_{\text{delivered}}(j) - Q^i_{\text{returned}}(j)) \tag{4.1}$$

where $\Delta Q^i(j)$ is the extra amount of material $i$ saved (if the amount is a positive value) or wasted (if the amount is a negative value) by group $j$; $Q^i_{\text{delivered}}(j)$ denotes the total quantity of material $i$ requested by group $j$; and $Q^i_{\text{estimated}}(j)$ denotes the estimated quantity that includes the statistic amount of normal wastage. The value of $Q^i_{\text{estimated}}(j)$ has to be carefully decided according to the circumstances of construction projects and previous experience (Schuette and Liska 1994; CIOB 1997). The $Q^i_{\text{returned}}(j)$ is the quantity of unused construction materials returned to the store by group $j$.

At the end of the project, the overall performance of group $j$ can be measured by Equation 4.2.

$$C^i(j) = \sum_n \Delta Q^i(j) \times P_i \tag{4.2}$$

where $C^i(j)$ denotes the total amount of material $i$ saved (if $C^i(j)$ is positive) or wasted (if $C^i(j)$ is negative) by group $j$; $n$ is the total number of tasks in the project that need to use material $i$; and $P_i$ is the unit price for material $i$.

The contracting company has to develop a policy to specify how the company shares the costs/benefits incurred from the reduction or increase of material wastes with workers. For example, the company may decide that workers should share 40% of the $C^i(j)$. In other words, the company will give back 40% of the $C^i(j)$ to workers as rewards. The rewards can be positive if the value of $C^i(j)$ is positive; and it can be negative (penalties) if the value of $C^i(j)$ is negative.

## 4.5  Implementation of IRP using bar-coding technology

### 4.5.1  Bar-code applications in construction

Since late 1980s, bar-code technology has been applied to many fields in construction as an automatic identification technology that streamlines identification and data collection on site. The application areas of bar-code technology in construction include quantity takeoff, field material control, warehouse inventory and maintenance, equipment/tool and consumable material issue, timekeeping and cost engineering, purchasing and accounting, scheduling, document control, office operations, and other information management in construction processes of projects (Stukhart and Pearce 1988; Stukhart and Pearce 1989; Stukhart and Cook 1989; Bernold 1990a,b; Stukhart and Cook 1990; Stukhart and Nomani 1992; McCullouch and Lueprasert 1994; Stukhart 1995; Bell and McCullouch 1998;

Chen and Li *et al.* 2000/2004). Some published studies regarding applications of bar-code technology in the construction industry are summarized in Table 4.6.

Although the bar-code technology has been used to control hazardous waste, including tracking information on hazardous material consumption and hazardous waste generation in the United States (Kemme 1998), no previous study has attempted to apply bar-code technology to minimize construction waste on sites before a crew IRP-based bar-code system was introduced (Li *et al.* 2003b). However, continued research of the crew IRP-based bar-code system conducted by the authors of this book shows that the proposed application is an efficient and

*Table 4.6* Research and applications of bar-code technology in construction

| Researcher | Year | Project | Field |
|---|---|---|---|
| Bell and Mc Cullouch | 1988 | Research | Potential applications |
| Stukhart et al.[a] | 1988/1995 | CII Research | Standardization |
| Lundberg and Beliveau | 1989 | Construction projects | Security management of M&E |
| Rasdorf and Herbert | 1989/1990a,b | Construction projects | Workforce and inventory management |
| Blakey | 1990 | Construction projects | Facility management |
| Bernold | 1990a,b | Testing | Construction environment |
| Brandon and Stadler | 1991 | Construction projects | Geotechnical data collection |
| Skibniewski and Wooldridge | 1992 | Construction projects | Robotic materials handling system |
| Baldwin et al. | 1994 | Precast concrete projects | Precast components management |
| McCullouch and Lueprasert | 1994 | Construction projects | Facility management |
| Stanley-Miller construction company | 1996 | Construction projects | Warehouse management |
| Echeverry et al.[b] | 1996/1998 | Construction projects | Personnel and materials management |
| Kemme | 1998 | Construction projects | Hazardous waste management |
| Wirt et al. | 1999 | Wastewater treatment plant | Equipment management |
| Li et al. | 2003b | Construction projects | Waste minimization |

Notes
[a] Stukhart and Pearce, 1988; Stukhart and Pearce 1989; Stukhart and Cook, 1989; Bernold 1990a,b; Stukhart and Cook 1990; Stukhart and Nomani 1992; McCullouch and Lueprasert 1994; Stukhart 1995; Bell and McCullouch 1998; Chen and Li *et al.* 2000/2004;
[b] Echeverry 1996; Echeverry and Beltran 1997; Echeverry *et al.* 1998.

cost-effective approach to integrating environmental management with project management in construction by implementing the crew-oriented IRP to minimize construction waste on sites.

### 4.5.2  Bar-coding system for IRP

As mentioned above, bar-code applications have been introduced to the construction industry since 1987 for material management, and plant and tool control (Bell and McCullouch 1988; Bernold 1990a,b; McCullouch and Lueprasert 1994; Stukhart 1994). The primary function of the bar-coding system is to provide instant and up-to-date information of quantities of materials exchanged between the store keeper and the group leaders/foremen. Specifically, implement IRP for reducing construction waste the bar-coding system can automatically

- track real-time data of new construction materials on the site;
- track real-time data of unused materials on the site;
- track real-time data of packing of materials and equipments;
- track real-time waste debris of materials on the site;
- record data of construction materials consumed in the project;
- monitor materials consumption of working groups;
- transfer real-time data to project management system;
- transfer real-time data of materials to head office via the Net.

The architecture of the bar-code system used in this implementation is illustrated in Figures 4.1 and 4.2. From these figures, it can be seen that when the group leader goes to the store to withdraw new materials or return surplus materials, the store keeper scans the bar-code labels for the materials as well as the bar-coding label/ID card of the group, so that the amounts of materials taken or returned by the group are registered in the database. Based on the amounts of materials initially ordered according to the estimated requirements, and the materials used by working groups, the computer system can calculate the value of $C^i(j)$ for each group $j$. Bar-codes are given to each item (if it is big, e.g. door, window, etc.) or each pack (if the items are small, e.g. pack of nails, bolts and nuts).

### 4.5.3  Material identification

For the materials, the bar-coding labels are designed to represent a material and its model, etc. For example, the code *0002-525-1-XYZ* represents "Cement – Portland, Ordinary 525# – 1 standard bag – XYZ Trademark", the code *0201-003-1-Local* represents "Aggregates – 3 mm particle diameter – 1 cubic meter – Local provenance", as shown in Figure 4.3. The "Class No." in Figure 4.3 is used to represent names of different materials, and the total number of the "Class No." is set as 2,000. The bar-code adopted for materials is Code 128 symbology (Stukhart 1995), and the codes are designed to represent Material, Model and

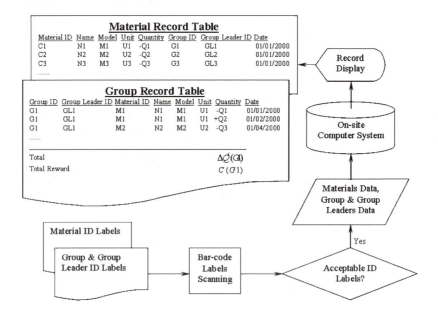

**Material Record Table**

| Material ID | Name | Model | Unit | Quantity | Group ID | Group Leader ID | Date |
|---|---|---|---|---|---|---|---|
| C1 | N1 | M1 | U1 | -Q1 | G1 | GL1 | 01/01/2000 |
| C2 | N2 | M2 | U2 | -Q2 | G2 | GL2 | 01/01/2000 |
| C3 | N3 | M3 | U3 | -Q3 | G3 | GL3 | 01/01/2000 |
| ...... | | | | | | | |

**Group Record Table**

| Group ID | Group Leader ID | Material ID | Name | Model | Unit | Quantity | Date |
|---|---|---|---|---|---|---|---|
| G1 | GL1 | M1 | N1 | M1 | U1 | -Q1 | 01/01/2000 |
| G1 | GL1 | M1 | N1 | M1 | U1 | +Q2 | 01/02/2000 |
| G1 | GL1 | M2 | N2 | M2 | U2 | -Q3 | 01/04/2000 |
| ...... | | | | | | | |

| Total | $\Delta Q(G1)$ |
|---|---|
| Total Reward | $C(G1)$ |

*Figure 4.1* Data flow diagram of the bar-code system for group-based IRP.

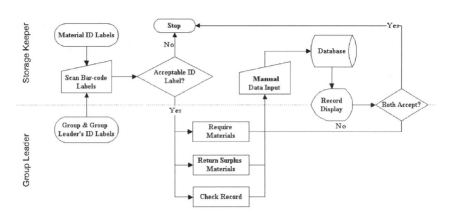

*Figure 4.2* Data flowchart of the bar-coding system for group-based IRP.

Quantity. For example, the code *0001-19-1* represents "plywood formwork – 19 mm thick – 1 square meters", as shown in Figure 4.3.

Because bar-code labels can be easily damaged during transportation and are cumbersome to scan if they are adhered onto the items/packs, we prepared a handbook of bar-code labels for all the construction materials used on site. This

# ABC CONSTRUCTION CO., LTD.
### Hung Hom, Kowloon, Hong Kong

## Construction Material Management System
## Bar-Code Lable

## Material Description

Name: Plywood formwork

Class No.: 0001

Model: 19 mm thick

Quantity Unit: 1 sq m

0001-19-1

Material ID No.: 0001-19-1

# ABC CONSTRUCTION CO., LTD.
### Hung Hom, Kowloon, Hong Kong

## Construction Material Management System
## Bar-Code Lable

## Material Description

Name: Cement

Class No.: 0002

Model: Portland, ordinary 525#

Quantity Unit: 1 bag

0002-525-1

Material ID No.: 0002-525-1

*Figure 4.3* Sample bar-coding labels for construction materials.

handbook contains all the bar-codes and is maintained and used by the material store keeper.

### 4.5.4  *Working-group identification*

For each working group, an identification card is issued to the group leader, who is responsible for collecting and returning construction materials. Figure 4.4 gives a sample identification card for a working group.

The bar-code of the group represents the group and its leader. For example, ID number *852-02-0100-017* represents "Carpenter group 852 and its leader's staff ID number is 02-0100-017", as shown in Figure 4.4. By scanning the bar-codes for the materials and the group, the computer system keeps records of materials

# ABC CONSTRUCTION CO., LTD.

### Hung Hom, Kowloon, Hong Kong

## Group Identification Card
## Bar-Code Lable

**Group Name: Carpenter**

**Type of Work: Formwork**

**Duty: Formwork**

**Group ID No.: 852-02-0100-017**

852-02-0100-017

**Group Leader: Henry G. Smith**

*Figure 4.4* Bar-coding label/ID card for a carpenter group.

used or returned by the group. These records are then used to calculate the reduction in or increase of material wastes generated by the group.

### 4.5.5  Hardware system

The hardware system of the bar-coding application consists of the bar-code scanner and the computer. A basic bar-code scanner consists of a scanner, a decoder, and a cable that interfaces between the decoder and the computer or terminal. Although there are four basic styles of bar-code scanners – light pen (usually called wand), linear CCD (charge-coupled device), laser, and video (CCD array) – the most versatile bar-code scanners are laser scanners, and many scanners have the decoder logic incorporated into a chip within the scanner, eliminating the need for a separate piece of hardware (PIPS 2001). The scanner we selected is PSC QuickScan 5385 scanner with keyboard wedge type of decoder integrated, which allows bar-code scanning to be added to almost any application without modification to the application software (PIPS 2001). Figure 4.5 illustrates the bar-coding hardware system.

### 4.5.6  Software system

The software system for a bar-code technology includes two essential software: bar-code–labelling software and bar-code–tracking software. Bar-code technology providers such as Loftware LLM-WIN32, BAR-ONE, and Bar-Tender, provide fast and easy-to-use bar-code–labelling software for designing

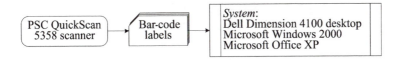

*Figure 4.5* Components of the bar-coding hardware system.

and printing quality labels. Bar-code–tracking software, such as IntelliTrack and Inventory Manager, can be used to read and track the bar-codes.

The bar-code adopted here is Code 128 symbology (Stukhart 1995). Software named "LLW-Win32 Design" (Version 5.x) from Loftware label printing systems is used to design the identification labels, and all bar-coding labels are printed out through a HP LaserJet printer. Identification of bar-coding labels is done using a handbook of bar-coding labels for all kinds of construction materials used on different sites, as discussed earlier.

### 4.5.7   Experimental results

A public housing project in Hong Kong was selected to experiment the group-based IRP. The project involved constructing two identical 34-storey residential blocks using a 6-day cycle. The 6-day cycle included nine major activities undertaken by nine working groups. The two blocks were constructed simultaneously by two teams of workers, each team having nine working groups with equal numbers of workers to carry out the 6-day-cycle construction method. We labelled the two teams as Team A and Team B. For the purpose of comparison, Team A did not adopt the group-based IRP during their operations, while Team B implemented the IRP with our advice and support.

The experiment has been conducted over three months. Results from Team's A and B during the three months are listed in Tables 4.7 and 4.8. The first column of the tables is the list of major materials used in the 6-day cycle. The second column is the unit of the materials; the third column contains the group names and their tasks. Columns 4–6 list estimated quantities of materials, quantities of materials delivered to groups, and quantities returned by groups. Column 8 lists the prices of materials, while columns 7–9 list results of calculations based on Equations 1 and 2. From the experimental results, it can be observed that throughout the three months, Team A consistently wasted more construction materials than Team B because workers in Team A did not see the benefits of reducing wastes. Therefore, by the end of three months, Team A had wasted additional amounts of construction materials valued at US$95,890.73 (HK$747,947.71). In contrast, Team B had made a substantial saving of US$90,428.83 (HK$705,344.85), indicating that the group-based IRP had effectively motivated workers in Team B in reducing avoidable wastes. The difference between the two projects is US$186,319.56 (HK$1,453,292.5). The cost of the bar-code system is

Table 4.7 Experimental results without group-based IRP (Team A)

| Materials | Unit | Group | | $Q^i_{estimated}(j)$ | $Q^i_{delivered}(j)$ | $Q^i_{returned}(j)$ | $\Delta Q^i(j)$ | $P_i$ | $C(j)$ |
|---|---|---|---|---|---|---|---|---|---|
| | | Name | Duty | | | | | | |
| Rebar | ton | Steel bender | Fix wall rebar | 1,760.00 | 1,795.20 | 0.00 | −35.20 | 2,271.31 | −79,950.11 |
| | | | Fix slab rebar | 1,408.00 | 1,425.60 | 0.00 | −17.60 | 2,271.31 | −39,975.06 |
| Precast façade | set | Rigger | Place precast façade | 1,760.00 | 1,760.00 | 0.00 | 0.00 | 3,000.00 | 0.00 |
| Precast slab | set | | Place precast slab | 9,856.00 | 9,856.00 | 0.00 | 0.00 | 1,500.00 | 0.00 |
| Cement | ton | Concreter | Concrete wall | 31,680.00 | 31,715.20 | 0.00 | −35.20 | 640.80 | −22,556.16 |
| | | | Concrete slab | 10,560.00 | 10,630.40 | 0.00 | −70.40 | 640.80 | −45,112.32 |
| | | Plasterer | Fit up wall, ceiling and floor | 15,400.00 | 15,554.00 | 0.00 | −154.00 | 640.80 | −98,683.20 |
| Sand | cubic meter | Concreter | Concrete wall | 26,928.00 | 27,280.00 | 0.00 | −352.00 | 57.04 | −20,078.08 |
| | | | Concrete slab | 10,560.00 | 11,264.00 | 0.00 | −704.00 | 57.04 | −40,156.16 |
| | | Plasterer | Fit up wall, ceiling and floor | 24,024.00 | 24,670.80 | 0.00 | −646.80 | 57.04 | −36,893.47 |
| Cobblestone | cubic meter | Concreter | Concrete wall | 26,752.00 | 27,456.00 | 0.00 | −704.00 | 58.30 | −41,043.20 |
| | | | Concrete slab | 10,560.00 | 11,264.00 | 0.00 | −704.00 | 58.30 | −41,043.20 |
| Hydrated lime | ton | Plasterer | Fit up wall, ceiling and floor | 9,394.00 | 9,424.80 | 0.00 | −30.80 | 464.00 | −14,291.20 |
| Plywood formwork | square meter | Carpenter | Fix timber slab form | 26,400.00 | 27,280.00 | 0.00 | −880.00 | 57.20 | −50,336.00 |
| Nail | bag | Rigger | Fix timber slab form | 1,760.00 | 2,640.00 | 0.00 | −880.00 | 50.10 | −44,088.00 |
| Drywall board | square meter | Rigger | Install wall board | 9,460.00 | 9,900.00 | 0.00 | −440.00 | 164.00 | −72,160.00 |
| Block | 10K blocks | Bricklayer | Bond masonry wall | 2.20 | 2.75 | 0.00 | −0.55 | 7,296.12 | −4,012.87 |
| Embedded plastic conduit | meter | Electrician | Conceal conduit installation | 18,480.00 | 22,000.00 | 0.00 | −3,520.00 | 1.05 | −3,696.00 |
| Glass | square meter | Glazier | Install window glass | 8,078.40 | 8,448.00 | 0.00 | −369.60 | 27.80 | −10,274.88 |
| Paint | square meter | Painter | Fit up minor works | 468.60 | 484.00 | 0.00 | −15.40 | 25.00 | −385.00 |
| Wall tail | square meter | Plasterer | Fit up wall | 22,704.00 | 23,760.00 | 0.00 | −1,056.00 | 34.00 | −35,904.00 |
| Mosaic | square meter | | Fit up wall and floor | 10,824.00 | 11,352.00 | 0.00 | −528.00 | 89.60 | −47,308.80 |
| | | | Total (HK$) | | | | | | −747,947.71 |

Table 4.8 Experimental results with group-based IRP (Team B)

| Materials | Unit | Group | Duty | $Q^i_{estimated}(j)$ | $Q^i_{delivered}(j)$ | $Q^i_{returned}(j)$ | $\Delta Q^i(j)$ | $P_i$ | $C(j)$ |
| | | Name | | | | | | | |
|---|---|---|---|---|---|---|---|---|---|
| Rebar | ton | Steel bender | Fix wall rebar | 1,760.00 | 1,724.80 | 17.60 | 52.80 | 2,271.31 | 119,925.17 |
| | | | Fix slab rebar | 1,408.00 | 1,372.80 | 17.60 | 52.80 | 2,271.31 | 119,925.17 |
| Precast façade | set | Rigger | Place precast façade | 1,760.00 | 1,760.00 | 0.00 | 0.00 | 3,000.00 | 0.00 |
| Precast slab | set | | Place precast slab | 9,856.00 | 9,856.00 | 0.00 | 0.00 | 1,500.00 | 0.00 |
| Cement | ton | Concretor | Concrete wall | 31,680.00 | 31,609.60 | 17.60 | 88.00 | 640.80 | 56,390.40 |
| | | | Concrete slab | 10,560.00 | 10,454.40 | 17.60 | 123.20 | 640.80 | 78,946.56 |
| | | Plasterer | Fit up wall, ceiling and floor | 15,400.00 | 15,276.80 | 15.40 | 138.60 | 640.80 | 88,814.88 |
| Sand | cubic meter | Concretor | Concrete wall | 26,928.00 | 26,576.00 | 105.60 | 457.60 | 57.04 | 26,101.50 |
| | | | Concrete slab | 10,560.00 | 10,384.00 | 211.20 | 387.20 | 57.04 | 22,085.89 |
| | | Plasterer | Fit up wall, ceiling and floor | 24,024.00 | 23,870.00 | 123.20 | 277.20 | 57.04 | 15,811.49 |
| Cobblestone | cubic meter | Concretor | Concrete wall | 26,752.00 | 26,576.00 | 246.40 | 422.40 | 58.30 | 24,625.92 |
| | | | Concrete slab | 10,560.00 | 10,384.00 | 211.20 | 387.20 | 58.30 | 22,573.76 |
| Hydrated lime | ton | Plasterer | Fit up wall, ceiling and floor | 9,394.00 | 9,332.40 | 3.08 | 64.68 | 464.00 | 30,011.52 |
| Plywood formwork | square meter | Carpenter | Fix timber slab form | 26,400.00 | 26,048.00 | 140.80 | 492.80 | 57.20 | 28,188.16 |
| Nail | bag | | Fix timber slab form | 1,760.00 | 1,584.00 | 88.00 | 264.00 | 50.10 | 13,226.40 |
| Drywall board | square meter | Rigger | Install wall board | 9,460.00 | 9,350.00 | 0.00 | 110.00 | 164.00 | 18,040.00 |
| Block | 10K blocks | Bricklayer | Bond masonry wall | 2.20 | 2.15 | 0.22 | 0.28 | 7,296.12 | 2,006.43 |
| Embedded plastic conduit | meter | Electrician | Conceal conduit installation | 18,480.00 | 18,004.80 | 176.00 | 651.20 | 1.05 | 683.76 |
| Glass | square meter | Glazier | Install window glass | 8,078.40 | 7,867.20 | 26.40 | 237.60 | 27.80 | 6,605.28 |
| Paint | square meter | Painter | Fit up minor works | 468.60 | 462.00 | 2.20 | 8.80 | 25.00 | 220.00 |
| Wall tail | square meter | Plasterer | Fit up wall | 22,704.00 | 22,545.60 | 132.00 | 290.40 | 34.00 | 9,873.60 |
| Mosaic | square meter | | Fit up wall and floor | 10,824.00 | 10,718.40 | 132.00 | 237.60 | 89.60 | 21,288.96 |
| | | | Total (HK$) | | | | | | 705,344.85 |

about HK$150,000. Thus, Team B has about HK$550,000 savings. These results convinced us that group-based IRP is effective in reducing construction wastes.

### 4.5.8   Crew IRP-based bar-code system

The crew IRP-based bar-code system comprises a crew-oriented IRP with a bar-code system (Li *et al.* 2003b). Previous research showed that the skill and attitude of workers are the main factors affecting the amounts of waste produced by workers (Pilcher 1992); between these two factors, their attitude towards work, including their enthusiasm and collectivism, is the most important in terms of waste generation. In addition, site surveys (Poon *et al.* 1996; Poon and Ng 1999) also indicated that workers' attitude towards construction operations and materials can make a significant difference to the amount of construction waste generated, and they may become careless in handling construction materials if there were lack of careful control and rewarding systems. As a result, reusable materials such as reinforcement bars, half-bags of cement, nails and timber pieces, etc. are often thrown away around the sites. The authors introduced the crew-based IRP thereafter in order to meet the demand of on-site construction material management. It is important to establish an on-site construction material management system to encourage workers to use materials carefully and efficiently, and to enhance their enthusiasm and collectivism by rewarding them according to their good performances in saving materials through reducing operational mistakes, returning unused materials for re-use or recycle, etc. (Li *et al.* 2003b). The crew IRP was conducted for on-site material management based on motivation theory by Maslow *et al.* (1998) and its development to CM such as the uses of special motivational programmes, and financial incentive programmes (Laufer and Jenkins 1982; Carberry 1996; Merchant 1997; Olomolaiye *et al.* 1998; Li *et al.* 2003b). It is expected that the crew IRP can help on-site CM to reduce any avoidable material waste caused by workers who may misuse materials on site.

As it is a quantitative approach to measuring the amount of material waste possibly generated in each construction process and each construction project, the computation of the crew IRP is done by using Equation 4.3.

$$C_i(j) = \sum_n \Delta Q_i(j) \times P_i = \sum_n (Q_i(j)_{es} - [Q_i(j)_{de} - Q_i(j)_{re}]) \times P_i \qquad (4.3)$$

where $C_i(j)$ is the total amount of material $i$ saved (if it is positive) or wasted (if it is negative) by crew $j$; $\Delta Q_i(j)$ is the extra amount of material $i$ saved (a positive value) or wasted (a negative value) by crew $j$; $P_i$ is the unit price for material $i$; $Q_i(j)_{es}$ is the estimated quantity that includes the statistic amount of normal wastage; $Q_i(j)_{de}$ is the total quantity of material $i$ requested by crew $j$;

$Q_i(j)_{re}$ is the quantity of unused construction materials returned to the storage by crew $j$; $i$ is number of any construction material that may be requested by a crew; $j$ is number of any construction crew whose operations may potentially generate waste; and $n$ is the total number of tasks in the project that need to use material $i$.

According to Equation 4.1, for a particular type of material $i$, the performance of crew $j$ in terms of material wastage can be measured by $\Delta Q_i(j)$, and at the end of the project, the overall performance of crew $j$ can be rewarded in agreement with $C_i(j)$. This means that the IRP is implemented according to the amount of materials saved or wasted by a crew i.e. if a crew save materials $(\Delta Q_i(j) > 0)$; the project manager will then award the crew a prize based on the amount of $C_i(j)$. In Equation 4.1, the value of $Q_i(j)_{es}$ has to be carefully decided according to the circumstances of construction projects and previous experience (Schuette and Liska 1994; CIOB 1997). On account of the requirement to increase the precision in reward through computation, a knowledge-driven system was introduced to re-use CM knowledge to more accurately define the value of $Q_i(j)_{es}$ (Chen and Li *et al.* 2005).

On the other hand, as construction waste is often generated due to misuse of materials by workers, the implementation of the crew IRP requires an efficient and cost-effective on-site material management system, and the bar-code system was thus adopted to implement the crew IRP (Li *et al.* 2003b). Figure 4.6 illustrates the architecture of the crew IRP-based bar-code system, which can be utilized on site in each construction project as mentioned with Site X in the figure.

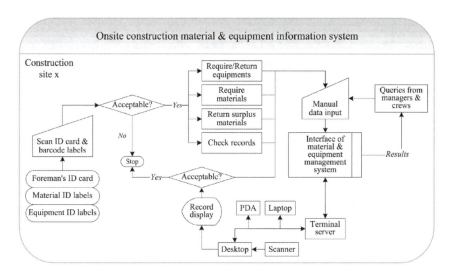

*Figure 4.6* A conceptual model for the crew IRP-based bar-code system.

The conceptual model described in Figure 4.7 comprises three sections: data capture mechanism, data process mechanism, and hardware system. Regarding the on-site M&E management, the data capture mechanism allows store keepers to scan bar-code labels of each M&E on site whilst facilitating crews and managers to input requests or queries related to M&E information. Meanwhile, the data process mechanism records the information of M&E and runs the IRP computation so that crews and managers can collect information for further decision-making on waste reduction. For example, bar-codes have been given to each M&E item and each pack; when a foreman goes to the store to request new materials or to return surplus materials, the store keeper scans bar-code labels corresponding to the materials as well as the ID number of the foreman, so as to collect information, such as the amount of materials taken or returned by the crew, for the M&E database. After the data collection, computations of IRP are done based on the amounts of materials initially requested by each crew but limited by the estimated quantities of each material, and the materials finally used by crews, and a software can calculate the value of $C_i(j)$ for each crew $j$. The value of $C_i(j)$ can thereafter be used to implement the IRP.

The hardware of the crew IRP-based system includes an on-site terminal computer server system, and immobile/mobile bar-code laser scanners. Table 4.9 gives an example of the hardware and software components of a crew IRP-based system application.

In this application example, bar-code representation adopted is the Code 128 symbology (Stukhart 1995), using Loftware® Label Manager to design the identification labels, and all bar-coding labels are printed out through a HP Laser-Jet printer. For each material and equipment, one bar-coding label is designed to represent one corresponding material or equipment and its model, etc.; for example, the code *0002-525-1-X* represents "Cement – Portland, Ordinary 525# – 1 standard bag – Trademark X", and the code *0201-003-1-Y* represents "Aggregates – 3 mm particle diameter – 1 cubic meter – Provenance Y" (Li *et al.* 2003b). One bar-coding label is designed to represent one crew; for example, coding number *586-01-0208-010* represents "Concretor crew 586 and its leader's staff ID number is 01-0208-010". By scanning the bar-codes for materials and crews,

*Table 4.9* An example of crew IPP-based system application

---

*Hardware*
Dell® Dimension® 4100 desktop
PSC QuickScan® 5385 scanner with keyboard wedge type of decoder
Handbook of bar-code labels for construction M&E (internal)

*Software*
Microsoft® Windows® NT/XP
Microsoft® Office® XP
Loftware® Label Manager

---

the computer system keeps record of materials used or returned by the crew. All these records are further used to calculate the possible wastes from each crew. Experimental results indicated that there is about 10% material saving by implementing the crew IRP-based bar-code system (Li *et al.* 2003b).

## 4.6 IRP and quality-time assurance

As the IRP focuses on waste reduction on site, the construction process might be jerrybuilt when a worker group wants to excessively save materials. It is important to integrate the IRP with quality and time management during the whole construction project. In the Hong Kong construction industry, residential buildings are built based on standard designs; it is convenient for the quantity surveyors to accurately measure the exact amounts of materials consumed in each activity and process. Working groups and the group foremen will be seriously questioned if the groups reduced material consumption in certain activities or processes such that the actual amounts of used materials were near or below the exact amounts measured by the quantity surveyors. In addition, rigorous quality assessment has to be conducted to ensure that the quality level is maintained, and working groups who can reach high quality requirement will also be awarded besides the reward from the IRP. On the other hand, the IRP could affect the duration little in each construction process if we apply information technology, e.g. bar-coding technology, in its implementation, instead of manual recording and calculation.

## 4.7 Integration with GIS and GPS

### 4.7.1 Potentials of the crew IRP-based bar-code system

Generally, urban development directly leads to the increase of construction and demolition waste. Since 1970s, governments, practitioners, and academics have been advancing gradually in pursuance of efficient and cost-effective environmental management to reduce construction waste worldwide (Chen and Li *et al.* 2000/2005); however, the total amount of construction waste is still out of control due to rapid urban development and lack of effective tools for CM. The statistic chart presented in Figure 4.7 reveals a remarkably bullish tendency of C&D waste generation in Hong Kong in 1986–2003 while several thousand tons of C&D waste was disposed of at landfills everyday on average (HKEPD 1998a,b,c/2004a,b). With worldwide perspectives to the construction industry, the issue of minimizing construction waste is being dealt with through process reengineering, technique innovation, and information technology by environment-concious construction sectors. For example, Fishbein (1998) and Coventry *et al.* (1999) established a set of construction-waste prevention strategies focusing on the effective coordination of materials management, including efficient purchase and ordering of materials; just-in-time delivery; careful storage and the use of materials to minimize loss, maximize re-use, prevention of undoing and redoing; reduction of packaging

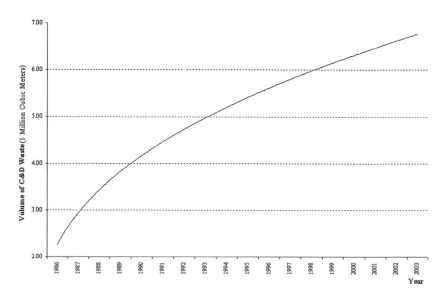

*Figure 4.7* The amount of C&D waste: a case in Hong Kong (1986/2003) (Data source: EPD, HK).

waste, etc. Previous studies on the construction-waste prevention strategies indicated that it is an extra expense for construction sectors to adopt new equipment and to utilize automation technologies in their projects (Ho 1997) and most (about 68–85%) construction sectors would adopt these new technologies only when it is requested by designers, specifications, or clients (Poon *et al.* 1996; Poon and Ng 1999), as a result, the cost-effective applications of information technology (IT) such as Web-based waste information exchange system (Chen and Li *et al.* 2003) can thus promote the deployment of the construction-waste prevention strategies.

Regarding IT applications in the area of construction-waste management, a crew-based IRP (Chen and Li *et al.* 2002a) with a bar-code system for on-site construction material management has been introduced to reduce any avoidable wastes by rewarding workers according to the amounts and values of materials they saved from their operations with the prerequisite of quality assurance. Compared with other IT applications for construction-waste management such as Web-based information exchange system about waste (Chen and Li *et al.* 2003), the IRP-based bar-code system can provide instant and up-to-date information about the quantities of materials requested or returned by a crew to a store keeper on site. Specifically, the bar-code system can automatically track real-time data of new materials, material residuals, material/equipment packing, and waste debris on the site.

However, there are two potentials of the on-site bar-code system. First, construction supervisors can comparably monitor the consumption of materials and equipment (M&E) in any similar ongoing construction processes and projects by using the recorded historical data of M&E utilized in any previous projects.

Second, construction managers and headquarters can re-use real-time information of M&E captured from each construction site in classified project management systems, including on-site construction M&E information system and central construction M&E information system. These potentials have left a research and development space for a more efficient CM information system to facilitate M&E management throughout the headquarters and each project on the platform of a wide area network (WAN). Considering this, the objective of this section is to present an integrated M&E management system using the IRP-based bar-code technology, the global positioning system (GPS) technology, the geographical information system (GIS) technology, and the WAN technology to facilitate M&E management, to control and reduce construction wastes, and to increase efficiency in project-oriented CM.

The methodology of the research comprises a combination of research methods including the development of an integrated physical model for M&E management in the enterprise-wide environment of construction sectors based on an extended literature review regarding the application of bar-code, GPS, GIS and WAN technologies in construction, and the adoption of the proposed model in a case study. Methods for achieving individual objectives are described below.

As mentioned above, potentials of former crew IRP-based bar-code system have left an opportunity to upgrade it from project-based M&E information system to enterprise-wide M&E management system by integrating GPS technology and GIS technology on the WAN, which is a geographically dispersed telecommunications network.

### 4.7.2 GPS/GIS applications in construction

The integrated utilization of GPS and GIS technologies is being adopted in more and more civilian areas to facilitate decision-making based on real-time remote-sensing spatial information. GIS is a computer-based system to collect, store, integrate, manipulate, analyse, and display data in a spatially referenced environment, which assists in analysing data visually and seeing patterns, trends, and relationships that might not be visible in tabular or written form (U.S.EPA 2004a,c). The application areas of GIS technology for environmental management include site remediation, natural resources management, waste management, groundwater modelling, environmental impact assessment, policy assessment compliance permit tracking, and vegetation mapping, etc. (U.S.EPA 2004a,c). On the other hand, GPS is a satellite-based navigation system made up of a network of approximately 24 satellites, which were placed into orbit by the U.S. Department of Defense in the 1970s and circle the earth twice a day in a very precise orbit and transmit information to earth, where GPS receivers receive this information and use triangulation to calculate the user's exact location (U.S.EPA 2004a,b). The application areas of GPS technology for civilian utilization include public safety, emergency location, automobile navigation, vehicle tracking, airport surveillance, control surveys, radial surveys, site acquisition and surveying, digital network timing and synchronization, precision farming, farm vehicle automation, and field environmental decision support, etc. (Bossler 2001; Kennedy 2002; U.S.EPA 2004a,b). In addition to the separated use of GPS technology or GIS technology in the mentioned areas, the integrated utilization of GPS and GIS technologies for civilian purposes also has increased since the 1990s (U.S.EPA 2004a,b,c; Hampton 2004).

In the fields of construction, both GPS technology and GIS technology, and their integrated technology have been introduced synchronously to many areas such as transportation management, facility delivery, urban planning, jobsite safety monitoring, site layout and development, and business analysis, etc. (Li *et al.* 2003a; Hampton 2004). Some published studies and applications of GPS and GIS technologies in the construction industry are summarized in Table 4.10. According to literature summarized in Table 4.10, the research and development of GPS/GIS applications in the construction industry was initiated in the early 1990s and there is still so much potential in the field of GPS/GIS applications

*Table 4.10* Research and applications of GPS/GIS technologies in construction

| Researcher | Year | System | Project | Field |
| --- | --- | --- | --- | --- |
| Selwood and Whiteside | 1992 | GIS | Civil engineering | Construction |
| Metcalf and Urban | 1992 | GIS | Highway corridor study | Highway construction |
| Bakken and Avey | 1992 | GIS | Water supply systems | Design and construction |
| Adams *et al.* | 1992 | GIS | Facility delivery | Urban planning |
| Williams | 1992 | GIS | Civil engineering | Construction |
| Jeljeli *et al.* | 1993 | GIS | Research | Contractor prequalification |
| Hammad *et al.* | 1993 | GIS | Bridge planning | Bridge construction |
| Launen | 1993 | GPS | Freeway monitoring | Transportation management |
| Parker and Stader | 1995 | GIS | Highway construction | Erosion predictions and control |
| Varghese and O'Connor | 1995 | GIS | Routing vehicles on sites | Construction planning |
| Issa | 1995 | GPS | Construction | Quality and productivity control |
| Robinson *et al.* | 1995 | GPS | Tunnel construction | Construction surveys |
| Nasland and Johnson | 1996 | GPS | Construction staking | Construction surveys |
| Cheng and O'Connor | 1996 | GIS | Site preparation | Construction planning |
| Udo-Inyang and Uzoije | 1997 | GIS | Highway construction | Inspection |
| Naresh and Jahren | 1997 | GPS | Vehicle tracking | Fleet management |
| Adams *et al.* | 2000 | GIS | Freeway monitoring | Oversize/weight permits |
| Wiegele | 2000 | GPS+GIS | Research | Pipeline construction |
| Cheng and Yang | 2001 | GIS | Site layout planning | Construction planning |
| Bernold | 2002 | GPS | Research | Construction engineering |
| Sacks *et al.* | 2003 | GPS | Labour monitoring | Workforce management |
| Li *et al.* | 2003a | GIS | E-Commerce | Material procurement |
| Sukut | 2003 | GPS | Heavy equipment control | Fleet management |
| McFall | 2004 | GIS | Sewer revision | Pipeline construction |

for construction sectors, comparing with the deployment of information systems in nearly all areas of construction engineering and management. In addition, most previous research and development focused on a single application of GPS technology or GIS technology, and the benefits of integrated GPS/GIS technology, which can bring highly efficient and cost-effective results to construction sectors, are still under excavation.

Although the integrated GPS/GIS technology has been used to provide decision-makers with the internal capability for rapid and effective contaminated site characterization (U.S.EPA 2004a,b,c), which is a typical utilization of the integrated technology in the area of environmental management to monitor and control adverse environmental impacts such as hazardous waste and noise, etc., there is no research initiative to apply the integrated technology to minimize adverse environmental impacts in construction such as construction waste and construction noise on sites. Since the integrated technology has been demonstrated to bring benefits in pipeline construction (Wiegele 2000), and either GPS technology or GIS technology can bring high efficiency and cost-effectiveness to construction sectors according to previous research and development (refer to Table 4.10), the authors further combine the crew IRP-based bar-code system with the integrated GPS/GIS technology to facilitate an enterprise-wide M&E management for the purpose of waste reduction. The proposed application will provide a highly efficient and cost-effective platform to assist the enterprise resource planning (ERP) implementation in the construction sector.

### 4.7.3   Integrated M&E management system

#### 4.7.3.1   Enterprise-wide crew IRP-based bar-code system

The enterprise-wide crew IRP-based bar-code system is a development of project-wide crew IRP-based bar-code system, which is presented in Figure 4.7. The aim of this development is to enhance the efficiency utilization of M&E information throughout the headquarters of a construction sector and each construction site belonging to it, from which the headquarters and site managers are able to get real-time information of M&E within the enterprise so as to make any further decisions depending on the information, such as the implementation of crew IRP in each project and the deployment of M&E among all projects. In addition, the enterprise-wide crew IRP-based bar-code system is an effective addition to a general-purpose construction project management system or an ERP system for construction sectors by means of automatic M&E data collection and data input through a terminal computer server on each construction site to a central computer server in the headquarters. Since Figure 4.7 has given an on-site section of the enterprise-wide crew IRP-based bar-code system, the central section of the proposed system is presented in Figure 4.8. Considering the possibility of M&E data input at the headquarters, the component of crew IRP-based bar-code system is combined to the central construction project management system, and

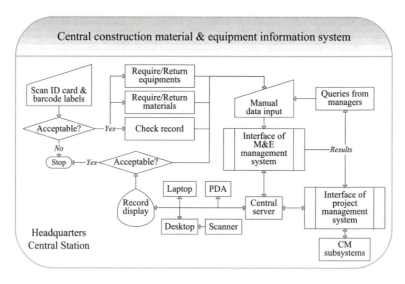

*Figure 4.8* A conceptual model for the enterprise-wide crew IRP-based bar-code system.

this system structure may facilitate central control without any obstacles such as authorization and firewall to go into any on-site M&E subsystems.

The data transfer among on-site M&E systems, central M&E system, and central construction project management system requires physical support from WAN. There are two main types of data transfer:

1 Data from construction sites regarding

- storage condition of M&E in each construction site,
- demand of M&E from each individual construction site,
- report of crew IRP from each construction site, and
- query and pivot of M&E to other construction sites and the headquarters; and

2 Data from headquarters, regarding

- query and pivot of M&E storage condition on each construction site,
- query and pivot of M&E demand from each construction site, and
- demands of M&E deployments from each construction site.

All this data transfer can be realized within a typical management information system, and the real-time communication between the headquarters and each construction site can be achieved based on the WAN. However, with the requirements of dynamic construction project management, the function and structure of traditional management information systems cannot provide satisfactory services regarding some real-time queries. For example, if managers from the headquarters want to know something about a kind of material, they may have questions about the present location of the material and its arrival time to a specific construction site (refer to Tables 4.11 and 4.12); but the limitation in information synchronization or real-time information capture in the traditional management information system necessitates an answer to these queries.

As a result, there is a requirement of plant information synchronization capacity for the traditional M&E management information system, and this capacity can make it easy to capture synchronous information from remote locations outside a construction site and the headquarters.

### 4.7.3.2 GPS/GIS integrated M&E management system

The integrated GPS/GIS technology adds new features such as construction vehicles tracking to the traditional M&E management information system for the propose of transferring real-time information about the location of any construction M&E that are being carried to a construction site from any locations outside the site. The integrated GPS/GIS technology helps to improve efficiency and increase profits by providing real-time vehicle locations and status reports, navigation assistance, drive speed and heading information, route history collection, etc. (Trimble 2004). Figure 4.9 illustrates the simple architecture of integrated GPS/GIS technology for the proposed M&E management system to reduce construction waste and to improve construction efficiency.

Regarding the cargo transportation of construction M&E, intercity freight transportation is dealt with in the proposed prototype (refer to Figure 4.9), including waterway transportation, air transportation, and overland transportation such as transportations by railroad and highway. Cargoes are fitted with GPS, which can transmit its positional data together with information about other attributes to the central station at the headquarters and distributed terminals on construction sites via the WAN. The central station at the headquarters is a monitoring station, where the accurate position of each construction cargo is displayed on a GIS map, and the information of each cargo can be queried. By using GIS analysis technology, the central station can get information about the current location of the cargo and estimate the time when the cargo can probably arrive at each predetermined construction site, i.e. its destination. Moreover, the central station also can send commands to drivers via personal digital assistants (PDAs) regarding cargo transportation and dispatch such as when they should start or which route they should pass through. This is very helpful for construction especially in a construction site where the space for material storage is limited; in theory, it is possible for zero storage on sites if the arrangement is precise and appropriate.

The deployment of the GPS/GIS integrated construction M&E management system requires physical support from computer hardware and software systems. The software requirements include computer operating system, GPS software, GIS software, and crew IRP-based M&E management system, etc. For example, a demonstration is developed in the Windows series of operating systems from Microsoft, including Microsoft Windows NT/2000/XP/CE and

Table 4.11 Real-time M&E information for supervisory control

| Material ID | Material name | Model | Unit | Quantity | Supplier | Provenance | Current location | Distance (km) | Demand time | Departure time | Arrival time |
|---|---|---|---|---|---|---|---|---|---|---|---|
| 0002-525-1-X | Cement | Portland, Ordinary 525# | Standard bag | x | X | HK | (x, y) | x.x | xx:xx xx/xx/xxxx | xx:xx xx/xx/xxxx | xx:xx xx/xx/xxxx |
| 0201-003-1-Y | Sand | 3 mm particle diameter | Cubic meter | x | Y | AU | (x, y) | x.x | xx:xx xx/xx/xxxx | xx:xx xx/xx/xxxx | xx:xx xx/xx/xxxx |

Table 4.12 Real-time M&E information for crew IRP

| Crew ID | Crew name | Material ID | Material name | Unit | $P_i$ | $Q_i(j)_{es}$ | $Q_i(j)_{de}$ | $Q_i(j)_{re}$ | $\Delta Q_i(j)$ | $C_i(j)$ | Current $Q_i(j)_{storage}$ | Demand time |
|---|---|---|---|---|---|---|---|---|---|---|---|---|
| 586-01-0208-010 | Concretor | 0002-525-1-X | Cement | Standard bag | x.xx | x.x | x.x | x.x | x.x | x.x | x.x | xx:xx xx/xx/xxxx |
| 586-01-0208-010 | Concretor | 0201-003-1-Y | Sand | Cubic meter | x.xx | x.x | x.x | x.x | x.x | x.x | x.x | xx:xx xx/xx/xxxx |

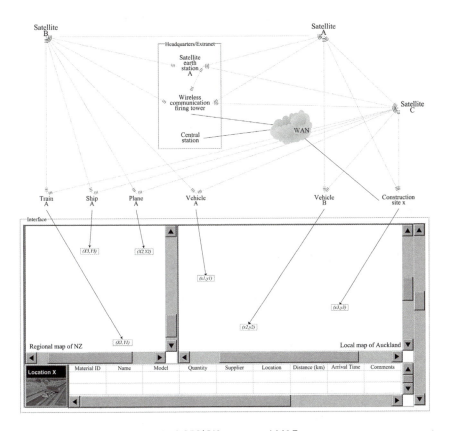

*Figure 4.9* A conceptual model of GPS/GIS integrated M&E management system.

Pocket PC, while the GIS software is ArcGIS series from ESRI (Environmental Systems Research Institute, of Redlands, California, USA), which is an integrated collection of GIS software products for building a complete GIS for organizations, including Desktop GIS (ArcReader, ArcView, ArcEditor, ArcInfo, and ArcGIS Desktop Extensions), Server GIS (ArcIMS, ArcGIS Server, ArcSDE, and GIS Portal Toolkit), Embedded GIS (ArcGIS Engine), and Mobile GIS (ArcPad, and Mobile ArcGIS Desktop Systems) (ESRI 2004); and the GPS software is GeoExplorer series from Trimble, which is an integrated collection of GIS-oriented GPS software products for advanced GPS/GIS data collection and mobile GIS tools, including Office Software (GPS Pathfinder Office and Trimble GPS Analyst extension for ArcGIS), and Field Software (TerraSync, GPScorrect for ESRI ArcPad, and GPS Pathfinder Tools Software Development Kit [SDK]) (Trimble 2004). The hardware requirements include enterprise-level computer server for control at central station, distributed computer desktop for operations on construction sites, and mobile laptop, Pocket PC, and handhelds for communications on the road. For

*Table 4.13* An example of GPS/GIS integrated M&E management system application

| Hardware |
| --- |
| Dell® Precision™ WS 370/670 (desktop)/ M60 (mobile) workstation |
| Dell® PowerEdge™ 6600 (enterprise-level) server |
| Trimble® GeoExplorer® series handhelds |
| PSC QuickScan® 5385 scanner with keyboard wedge type of decoder |
| Handbook of bar-code labels for construction M&E (internal) |
| *Software* |
| Microsoft® Windows® Server 2003 SE |
| Microsoft® Windows® NT/2000/XP/CE |
| Microsoft® Pocket PC |
| Microsoft® Office® XP |
| Loftware® Label Manager |
| ESRI® ArcGIS® series |
| Trimble® GeoExplorer® series |

example, Trimble GeoExplorer series handhelds, which are the most advanced GPS/GIS data collection and mobile GIS tools available, combining a Trimble GPS receiver with a handheld computer running Microsoft Windows Mobile 2003 software for Pocket PCs (Trimble 2004), was chosen to support the operation of the GPS/GIS data collection, including GeoXT, GeoXM, Beacon-on-a-Belt, External Patch Antenna, etc., whilst Dell PowerEdge enterprise-level server was chosen to operate the GPS/GIS integrated central construction M&E management system at central station, and Dell Precision series of desktop/laptop workstations were chosen to operate the GPS/GIS integrated on-site construction M&E management system on construction sites. Table 4.13 gives an example of the hardware and software components for the GPS/GIS integrated construction M&E management system application.

Provided the development period of the proposed GPS/GIS integrated construction M&E management system is not acceptable to an urgent need from a construction sector, commercial solutions such as the Trimble construction solutions (Trimble 2004) can be used.

### 4.7.4   A pilot study

#### 4.7.4.1   The problem

C&D waste has been identified as a priority waste in the New Zealand Waste Strategy because of its quantity and complexity, which sets a target of 50% waste reduction in waste being disposed of to landfills by 2008, and requires local authorities to put in place programmes for monitoring C&D waste quantities (MFE 2004). Under this circumstance, a Hong Kong-based construction company had an ongoing project at Auckland, New Zealand (refer to Figure 4.9), and the managers at the headquarters in Hong Kong wanted to deploy all construction M&E by using the GPS/GIS integrated construction project management system, and implement the crew IRP on the construction site at Auckland to fulfill the company's environmental promise in minimizing construction waste.

Regarding the construction M&E required in this project, most of them are supplied and transported from Australia and China, except a limited quantity of M&E which are ordered from local suppliers in New Zealand. In order to carry out the construction schedule on time and reduce waste, site managers at Auckland had to try hard to pay attention to their M&E management, and struggled with adverse atmospheric conditions in New Zealand. As a result, they asked the headquarters to provide much accurate information regarding the arrival time of construction M&E so as to deal with limitations of M&E storage on site and the varied weather conditions there.

### 4.7.4.2   Requirements specification

As managers from both the headquarters and construction site need dynamic accurate location information of M&E to push on their jobs effectively, the demand for immediate response time and tight command and control necessitates the GPS/GIS integrated solution to enable real-time interactive communications for dispatch and navigation, and server-based cargo tracking and messaging and others. A construction fleet management process based on the GPS/GIS integrated construction M&E management system (refer to Figure 4.9) should have the following major positioning-related requirements from both the headquarters at Hong Kong and the construction site at Auckland:

- Efficient dispatch and supervisory central control of cargoes among the construction site and the M&E suppliers from China, Australia, New Zealand, and other places at the headquarters side, which means

    1   to correctly arrange the departure time and routes of each cargo from suppliers,
    2   to accurately define the arrival time of each cargo at the construction site,
    3   to actively track the dynamic position of each transportation,
    4   to timely monitor and control the process of each transportation from departure to arrival, and
    5   to dynamically record any delay due to the transportation by suppliers for further claiming indemnity, etc.

- Efficient dispatch and supervisory on-site control of cargoes on the road and M&E on site at the construction-site side, which means

    1   to dynamically check the location of each cargo on the road to the construction site,
    2   to timely communicate with the headquarters about each transportation,

3   to accurately record the arrival time of each cargo on site,
4   to accurately record the details of each cargo arrived on site, and
5   to accurately record the details of each material or equipment received by each crew, etc.

### 4.7.4.3   Solutions

The headquarters at Hong Kong chose the outlined commercial solutions from Microsoft, Loftware, ESRI, and Trimble (refer to Table 4.13) to provide integrated GPS/GIS capabilities for managers on both sides for dynamic construction M&E management. There are two phases in the deployment of the application. In Phase I, proposed GPS/GIS devices including software system and hardware system are planted into the currently used construction M&E management system, which is integrated as a M&E subsystem with an enterprise-wide construction project management system. Detailed GPS coordinate information, including extensive map as well as latitude, longitude, date, and time, could be displayed in the system. The enhanced system was modified to interface with the headquarters' existing construction project management system and construction engineering system. In Phase II, Trimble external patch antenna (EPA) is adhered to each cargo on the road to the construction site as a kind of vehicle location device (VLD). The Trimble EPA is specially designed for seamless integration with Trimble GeoExplorer series handhelds and the WAN infrastructure, and is ideal for use in all environments where a high yield of positions is required (Trimble 2004). By automatically positioning each transportation, real-time information of each cargo will be accessible for both central and on-site construction M&E management systems.

### 4.7.4.4   Results

Managers in both the headquarters at Hong Kong and the construction site at Auckland were satisfied with the novel application of integrated GPS/GIS technology in construction management, in which the specified values are all actually achieved, such as the reduction of construction waste and the improvement of construction efficiency. Table 4.14 provides a comparison of the non-integrated system versus the GPS/GIS integrated system for the construction M&E management solution. According to the comparison, the GPS/GIS integrated solution can improve the construction efficiency through increasing the effective working hours of construction equipment and reducing construction duration and the cost of workforce, as well as reduce the generation of construction waste. Due to the initial investment in the hardware and software systems, original cost increased compared with the former crew IRP-based bar-code system; however, this can be overcome during further utilization of the new system.

*Table 4.14* Comparison of non-integrated system versus the GPS/GIS integrated system

| Parameter | Non-integrated system | Integrated system | Variation (%) | Compliant |
|---|---|---|---|---|
| Hardware cost | $2,500 | $8,000 | ↑220 | No |
| Software cost | $1,200 | $6,000 | ↑400 | No |
| Equipments utility | 3,100 hours | 3,600 hours* | ↑16 | Yes |
| Construction duration | 210 days | 195 days* | ↓7 | Yes |
| Workforce cost | $400,000 | $360,000* | ↓10 | Yes |
| Construction waste | $8,500 | $2,000* | ↓77 | Yes |
| Cost–benefit integration | | | | Yes |

* Denotes predicted values.

### 4.7.5 Conclusions and recommendations

This section aims to enhance the crew IRP-based bar-code system for construction M&E management by utilizing integrated GPS/GIS technology. By integrating GIS/GPS with the crew IRP-based bar-code system, real-time information on location, quantities and types of construction materials can be effectively tracked. In order to achieve this objective, the former project-oriented crew IRP-based bar-code system was first extended to an enterprise-wide construction M&E management system which was integrated to the traditional construction project management system. The extended prototype was further developed to a GPS/GIS-integrated construction M&E management system, as managers in both headquarters and construction sites have the need to get real-time information to control cargoes on the road to sites and to reduce waste generation on sites. The authors then present the conceptual model for the proposed GPS/GIS-integrated system with its logical system design and system implementation. Potential requirements and further applications are discussed as well. Finally, a case study is done to demonstrate the cost benefit of the novel system in construction. It is expected that the proposed innovation, which changes the M&E management from process-focused partial waste prevention to project-oriented total waste reduction, can dramatically improve the serviceability of the bar-code system in real-time data capture and re-use to assist the ERP implementation of construction sectors.

## 4.8 Conclusions and discussions

Although experimental results demonstrated the obvious strength of the group-based IRP in reducing wastage of construction materials, there has been a concern from the senior management of the contracting company in using the group-based IRP. The concern was the fear that workers might jerry-build in order to save materials, as the IRP does not directly relate itself to the quality of work. Therefore, the management felt that there is a need to investigate how to combine the quality and time performances of workers with the IRP when deciding the amount of rewards to workers. The IRP-integrated construction management has been proved to be useful and effective in the implementation of IRP.

Difficulties have also been identified during implementing the IRP on site. First, because the bar-code system can only recognize materials that have the standard quantity and does not automatically accept returned bits and pieces, quantities of the returned materials have to be assessed by the store keeper and be manually entered into the computer. This can potentially bring inaccuracies into the system. Second, as different groups may withdraw same materials, misunderstanding and conflicts between groups may occur if materials of one group are moved or mistakenly used by members of other groups. This problem will be intensified in situations with congested working spaces. These problems need to be resolved before the group-based IRP can be fully accepted and endorsed by the industry.

This chapter presents a group-based IRP, which encourages workers to reduce avoidable wastes of construction materials on site. The IRP is based on the principle of motivating workers through giving them performance-based financial rewards. Because of the unique situation in Hong Kong, this study did not consider other factors that may influence the generation of on-site wastes, such as design coordination and site supervision. Therefore, further studies are needed to test the usability of the IRP in other countries. In addition, this chapter introduces the use of a bar-code system to register the flow of materials so that performances of working groups in terms of material wastage can be easily measured. In order to avoid jerry-building, further research is needed to integrate the IRP with quality and time management.

This chapter also uses an integrated GPS/GIS technology in the reduction of construction waste and the increase of efficiency in project-oriented construction management. There are two relevant sections to describe the application in this chapter. First, a system prototype is developed from an automatic data capture system such as the bar-coding system for construction M&E management onsite, whilst the integrated GPS/GIS technology is combined with the M&E system based on the WAN. Second, a case study is done to demonstrate the deployment of the proposed application. Besides the presentation of the conceptual model, the logical system design, and the system implementation of the integrated M&E system, it is expected that the proposed innovation, which enhances the M&E management from process-focused partial waste prevention to project-oriented total waste reduction, can dramatically improve the serviceability of the bar-coding system in real-time data capture and re-use to assist with the ERP implementation of the construction sector.

# Chapter 5

# Effective reduction at post-construction stage

## 5.1 Introduction

As a modern way to conduct business in the global economic environment, e-commerce is becoming an essential component integrated with traditional business processes in enterprises. In order to reduce risks and increase profits in e-commerce investments and provide the best services to their customers, enterprises have to find appropriate approaches to analyse their e-commerce strategies at business planning stage. Strategic management tools are designed for enterprises to evaluate their business strategies and they can be used to evaluate the e-commerce business plan as well. For example, the SWOT (strengths, weaknesses, opportunities, and threats) analysis is regarded as a popular way to evaluate an e-commerce business plan, with business environmental scanning based on internal environmental factors (strengths and weaknesses) and external environmental factors (opportunities and threats) (Turban *et al.* 2003). In order to facilitate the application of the strategic management tools, different forms of applications are adopted, such as checklist (OGC 2004), rating system (UNMFS 2004), and expert system (PlanWare 2004), etc. Among these strategic management tools, computer-driven business simulation tools enable participants to run virtual business processes, experiment with different strategies, and compete with other companies or plans in a virtual business environment. As an example, the Marketplace (ILS 2003; IDC 2004) is a business simulator for integrative business courses, which provides decision-making content on marketing, product development, sales force management, financial analysis, accounting, manufacturing, and quality management. Regarding the application of computer simulation in e-commerce, the Marketplace strategic e-commerce simulation is designed specifically for e-commerce courses, and it illustrates the business concepts of the e-commerce environment (ILS 2003). For an e-commerce system simulation, Griss and Letsinger (2000) conducted research on flexible, agent-based e-commerce systems with an experimental multi-player shopping game, in which agents represent buyers, sellers, brokers, and services of various kinds, for demonstration and educational value, for experimenting with alternative individual and group economic strategies, and for evaluating the effectiveness of agent-based systems for e-commerce. Both academic and professional practice have proved that using computer-driven simulation is an effective, efficient, and economical way for e-commerce business plan evaluation.

However, it is hard to conduct simulation based on the detail flowchart of business processes within the current e-commerce simulation environment as mentioned above. This actually limits the application of e-commerce simulation. In fact, computer simulation has been applied to tackle a range of business problems, leading to improvements in efficiency, reduced costs, and increased profitability since the 1950s (Robinson 1994). During this period, the use of simulation software tools was on the rise in various application areas (Google 2005) and process-oriented simulation had become popular in business management (Swain 2001). The authors believe that a process-oriented simulation for e-commerce system evaluation is more directly perceived through the human sense, and their interest is to conduct a quantitative approach to e-commerce system evaluation based on the theory of process simulation.

The e-commerce system simulation is an integrative procedure to run a business-process-oriented simulation programme based on both internal and external business environmental factors to demonstrate the actual results of implementing an e-commerce business model by using computer-driven software toolkits. The e-commerce system simulation is an effective, efficient, and economical approach, and can be used to experiment e-commerce business models and to evaluate different e-commerce business plans, in which quantitative analysis is required by decision-makers. The adoption of e-commerce system simulation can overcome some limitations in e-commerce system development such as the huge amount of initial investments of time and money, and the long duration from business planning to system development, then to system test and operation, and finally to exact returns; in other words, the proposed process oriented e-commerce system simulation can help currently used system analysis and development methods to tell investors in a very detailed way about some details of keen interest such as how good their e-commerce system could be, how many investment repayments they could have, and which area they should improve from initial business plans.

The definition of the e-commerce system simulation has actually normalized a procedure to apply process simulation to run an e-commerce model at system-design stage. In this regard, this chapter will focus on the adaptation of an e-commerce model into a process simulation environment. And the authors achieve this through experimental case studies with an e-commerce business plan, called Webfill, for online C&D waste exchange in Hong Kong. The methodologies adopted in this chapter are literature review, system analysis and development, simulation modelling and analysis, and case study. Results from this chapter include the conception of e-commerce system simulation, a comprehensive review of simulation methods adopted in e-commerce system evaluation, and a real case study of applying simulation to e-commerce system evaluation. Furthermore, the authors hope that the adoption and implementation of the process simulation approach can effectively support business decision-making and improve the efficiency of e-commerce systems.

## 5.2 Background

Generally speaking, C&D waste can be reduced by using innovative construction techniques and management methods, such as adopting prefabrication and installation technologies, recycling C&D debris, reducing the possibility of waste generation in architecture and structure design, and improving site-based materials management, etc. Although these approaches have proved to be effective to some extent, most of them are still in a stage of research, and contractors usually do not like to invest in high-cost techniques and approaches if they were not forced to do so. For example, surveys show that local constructors in Hong Kong feel it is expensive to use new machinery and automation technology (Ho 1997); most (68–85%) local constructors agree to adopt alternative low-waste but high-cost techniques only when they are demanded by the designers, the specifications, or the clients (Poon and Ng 1999). As a result, C&D wastes are normally not controlled effectively on construction and demolition sites in Hong Kong. According to statistical data, C&D debris frequently makes up 10–30% of the waste received at many landfill sites around the world (Fishbein 1998), but this figure has been over 40% in recent years in Hong Kong (refer to Table 5.1).

In contrast to the percentage in other advanced countries, for example, C&D debris makes up only 12% of the total waste received at Metro Park East Sanitary Landfill of Iowa State in the United States (MWA 2000); the quantity of C&D waste in Hong Kong is about three to four times higher. So there is an urgent need to deal with the problem and to find a practical solution for C&D waste reduction in Hong Kong.

*Table 5.1* An analysis of C&D waste disposal in Hong Kong (HKEPD 1999a,b,c,d/2004a,b)

| Year | Amount of waste disposal at landfills (ton/day) | | Percentage of C&D waste (%) |
|---|---|---|---|
| | C&D waste | Total waste | |
| 1998 | 7,030 | 16,738 | 42 |
| 1999 | 7,890 | 17,932 | 44 |
| 2000 | 7,470 | 17,786 | 42 |
| 2001 | 6,410 | 16,686 | 38 |
| 2002 | 10,202 | 21,158 | 48 |
| 2003 | 6,728 | 17,757 | 38 |
| Average | 7,621 | 18,010 | 42 |

One of the most important C&D waste control regulations in Hong Kong is the trip-ticket system (TTS) for disposing waste from work sites to disposal facilities and landfills, which was originally recommended in the *Waste Disposal Ordinance & Waste Disposal (Chemical Waste) (General) Regulation* in Hong Kong in 1998, and was formally adopted in the Hong Kong construction industry on July 1, 1999 (HKEPD 1999a,b,c,d). The aim is to control illegal dumping and ensure proper disposal of C&D waste at public filling facilities or landfills. The TTS is a system for recording orderly disposal of C&D waste to disposal facilities by trucks. Under the TTS, contractors are required to fill in a standard trip-ticket form outlining the details of the transportation vehicle, type and approximate volume of waste, and the designated disposal facility which has been approved by the Public Fill Committee or the Director of Environmental Protection of the Government (CED 2002). Once the C&D waste is delivered to the designated facility, a receipt is issued to the vehicle operator for returning to the project engineer or architect representative for verification of the contractor's compliance with the policy requirements, and the contractors are then charged based on their receipts by the disposal facilities. The TTS is implemented to ensure a certain level of accountability among the project proponent, engineer/architect, and the contractor. Moreover, it facilitates the recording of waste as it arrives at the landfill or public filling area and minimizes the potential for cross-contamination with other waste which the vehicle operator may otherwise likely pick-up and route to the disposal facility. The TTS assumes that the contractor will bear the responsibility for the sorting (where applicable) of the C&D material generated on their site prior to its disposal.

According to the environmental permit conditions to construct and operate a designated project in the *Hong Kong Environmental Impact Assessment Ordinance*, the disposal of C&D waste should be controlled through the TTS and the records should be readily available at all times for inspection at all site office(s) covered by the Environmental Permit (HKEPD 2000a,b,c). As a result, hundreds of public-works project contracts and Housing Authority contracts invited have applied the TTS following their environmental permits in Hong Kong, and each of them obtained an admission ticket from the Facilities Management Group of the EPD for disposal of contaminated soil at landfills. From the environmental impact reports submitted recently by contractors in Hong Kong (HKEPD 2002a,b,c), the TTS is used to audit C&D waste disposal records to ensure that the number of vehicles/trucks leaving the construction site corresponds with the number of deliveries at the landfills. An on-site environmental team is normally set as an independent checker to audit the implementation of the TTS and ensure proper disposal and avoidance of fly tipping.

On the other hand, generally, the development of real estate in urban areas directly leads to the increase of C&D waste; so a great deal of efforts have been made by both academics and professionals to reduce on-site waste during construction. Although governmental audiences and industrial practice to reduce C&D waste are ordinarily known and construction contractors are encouraged

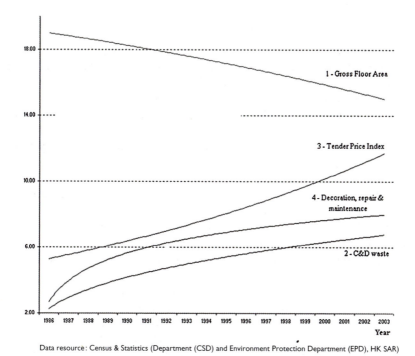

Data resource: Census & Statistics (Department (CSD) and Environment Protection Department (EPD), HK SAR)

*Figure 5.1* A statistic chart of C&D waste and real-estate development in HK.

to apply environmental management on site, the reason why C&D waste keeps increasing has not been made clear. However, a statistic analysis as presented in Figure 5.1 reveals that there is a remarkable divergence between the bullish tendency of C&D waste generation and the bearish tendency of real-estate development in Hong Kong since the mid-1980s, and this evidence indicates that the scale of construction in real-estate development may not play a leading role in generating C&D waste. On the other hand, the statistic analysis also reveals that synchronous tendencies exist between trend-line 2 and trend-line 4 in Figure 5.1, which indicates that building decoration, repair, and maintenance works are a real leader in generating C&D waste in Hong Kong. As the authors did not extend this statistical analysis outside the Hong Kong construction industry, it is not rational to conclude that most part of C&D waste is generated due to renovation works of buildings in a worldwide scale; however, the statistic analysis emphasizes the importance of considering property management activities at post-construction stages in the C&D waste exchange process.

The Government of the Hong Kong Special Administrative Region (HKSAR) has proposed to implement a C&D waste management strategy in the *Government Plan 1999–2007*, which is essentially to avoid, minimize, recycle, and dispose of waste based on desirability. The target of the strategy is to reduce the generation of C&D waste and hence its intake at landfills, and to re-use and recycle as much C&D material as possible. Similar to the *C&D Debris Management Program* that has been put into practice at the Metro Park East Sanitary Landfill site since 1995 in the United States (MWA 2000), tipping fee (HK$125 [about US$16] per ton) on C&D waste taken to landfills is imposed in Hong Kong since 1999, when the Environmental Protection Department (EPD) of the HKSAR government established an administrative TTS in public-works project contracts for the proper disposal of C&D waste at public filling facilities or landfills (HKEPD 1999a,b,c,d). Benefits of implementing this strategy include potential savings for landfill sites, proper disposal of C&D waste, and reduction of waste generation on site. However, it is reported that the TTS encounters obstacles when waste transporters are asked to pay disposal charges for contractors who are the generators of the C&D waste, and transporters are finally permitted legally to dispose the waste without any payment if they can make a written or even an oral pledge that contractors have not paid for them (Mingpao 2002). This legal loophole indicates it is necessary to improve the TTS through better managing the flow of waste disposal.

Although the authors have not found any report on how much C&D waste has been recorded and how much C&D waste has been reduced due to the implementation of the TTS in Hong Kong, it is not difficult to find out that the TTS's contribution is limited in the whole C&D waste cycle. The main feature of a C&D waste cycle with a smooth movement and operation is that it must be a valued-added chain, where all participants including construction contractors, property managers, material manufacturers, waste material recyclers, landfill managers, etc. can get benefits. However, the TTS can only record about waste conveyance between construction sites and landfills, and it seems to have no direct contribution to the value-added chain, even to the reduction of C&D waste. Specifically, main weaknesses of the TTS exist in the following four aspects:

1  current TTS is only implemented in public construction projects, and the disposal of the C&D waste generated from private construction projects is not controlled;
2  although the TTS tracks the results of C&D waste disposal, information of waste tracking is not used effectively in waste management;
3  tipping fee of C&D waste can be waived as an expedient; and
4  the TTS increases the amount of paperwork.

The EPD of the HKSAR government has been attempting to resolve the last two weaknesses by introducing new legislations and a smart card system. This chapter focuses on introducing an e-commerce system called Webfill, which is an online portal for C&D waste trade, so that all participants can benefit from using this system.

In order to deal with these problems in C&D waste management in Hong Kong, this chapter proposes to apply an e-commerce model for C&D waste exchange to enhance efficiency and effectiveness of the TTS and accordingly to reduce the total amount of C&D waste disposed to landfills in Hong Kong. For fear that the e-commerce model would not provide an ideal result, a simulation-based comparison between the existing TTS and the enhanced TTS is conducted. With the only view of reducing the C&D waste in Hong Kong, the e-commerce model or the waste exchange model can only work for reducing the already generated C&D waste, while generation of the waste cannot be expected to be controlled with it. As a result, this chapter only focuses on the e-commerce model for the C&D waste reduction on a post-construction stage.

## 5.3 Online waste exchange approach

### 5.3.1 Feature comparison of waste-exchange websites

The concept of waste exchange systems for exchanging industrial residues and information, and for reducing the waste volume was introduced in the 1970s (Middleton and Stenburg 1972; Mueller *et al.* 1975). In recent years, Web-based services for waste material and equipment trade and information exchange have been developed as they support effective multimedia communication. Online search results show that there are a number of websites related to waste exchange, and some of them also provide in advance a special area for quality salvaged C&D waste at comfortable prices on their websites; however, it is apparent that no website has been found to be solely dedicated to e-commerce of C&D waste exchange. For a website review, Appendix C summarizes 36 online C&D waste-exchange-related websites. According to relevant information from these websites, the authors made a short comparison study based on criteria described in Table 5.2, and Figure 5.2 shows a statistic comparison of these websites.

Based on the website review, the authors noticed that there is a general online C&D waste exchange model, adopted by most of the observed websites, which

*Table 5.2* Feature comparison of C&D waste exchange websites

| Website name | Market | Functions | | | | | Charge | Condition remarks |
|---|---|---|---|---|---|---|---|---|
| | | Search | List | Add data | Trade online | Membership | | |
| Website 1 | Local/Global | √/x | √/x | √/x | √/x | √/x | Free/Not | Working/Not |
| Website 2 | Local/Global | √/x | √/x | √/x | √/x | √/x | Free/Not | Working/Not |
| . . . | | | | | | | | |
| . . . | | | | | | | | |
| Website *n* | Local/Global | √/x | √/x | √/x | √/x | √/x | Free/Not | Working/Not |

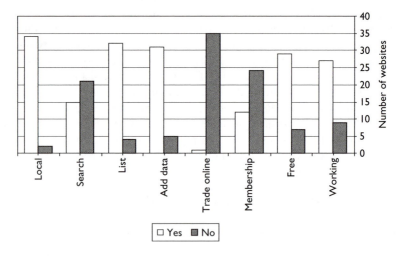

*Figure 5.2* Feature comparison of C&D waste exchange websites.

can be developed based on their common features summarized in Figure 5.2. Although there are some differences among their profile designs, common features exist in data transfer and website functionalities.

### 5.3.2  Operation obstacles

The websites for waste exchange have generally proved to be useful and effective in reducing total industrial waste. For example, since 1992 more than 650,000 tons (Note: this figure remains unchanged in 2002) of materials have been diverted from landfills and over 5.5 million dollars have been saved through the CMX (CMX 2000). However, it has also been found that information about C&D waste or the number of contractors who want to buy second-hand materials or C&D waste is very limited. For instance, search results of the CMX show that there are no date records about every available material and there is no buyer requesting C&D waste materials (CMX 2000), and this situation recurs in other observed websites (accessed between 2000 and 2003), e.g. HappyHarry's, HIMAX, and MaterialsExchange, etc. To quote examples for the status of online C&D waste exchange, HappyHarry's provides for used building materials exchange on the Internet; although there are about 51 records on the webpage under item "View list", record numbers are only available from 970728-1 to 971205-2, and these numbers indicate that there might be no recent records, and it has not worked effectively since 1998. The HIMAX was in a similar situation, where only records of 1996 can be found; and MaterialsExchange is another website providing services for C&D waste exchange, where similarly, only two records were found on its material list, and the records were last revised on

April 11, 1999. From these observations, it is reasonable to assume that websites for waste exchange are not widely accepted in the construction industry, and the main suspected reasons are given below:

1    contractors pay less attention to C&D waste reduction;
2    contractors can make little profit by using waste exchange;
3    information of waste exchange is scattered on many different websites; and
4    websites lack user-friendly/efficient operational mechanism's to pull users.

The main problem of subsidence of online C&D waste exchange comes from contractors, who pay less attention to C&D waste reduction. Over a long period of time, contractors have been accustomed to conventional project management, including cost management, time management, and quality management, and environmental management during project construction is relatively new to them. In many developing countries, contractors are still allowed to transport their C&D waste to landfills for free, rather than using a Web-based tool to find the best ways of recycling C&D waste and which delivers customer requests directly to contractors' desks.

Another problem is that contractors make little profit from using waste exchange systems. In many parts of the world, contractors have to pay for disposing of C&D waste to landfill sites, and contractors are being pressured to reduce the C&D waste discharge. Under this circumstance, a Web-based C&D waste exchange site becomes necessary to contractors as the website can disseminate information about C&D material which could be reusable by other people. According to statistics from the Portland area in the USA, there were approximately 550,000 tons of C&D wastes (about 145,000 drop box loads) in 1994. While garbage-dumping fees are US$62.50 per ton, over 50% of the C&D waste can be diverted to a recycler (buyer) for incomes ranging from nothing to US$35 per ton (Metro 1997). By using an online C&D waste exchange system, contractors can sell their residual materials to other contractors or manufacturers or recyclers to reduce their C&D waste disposal costs and conserve resources. However, the current Web-based information exchange model only provides contractors an information-exchange platform. No matter whether they want to sell or buy, contractors and manufacturers will have to wait with patience for feedback information from each other, and this often leads to delays in construction or manufacture processes. So contractors often have to give up the benefit from selling out their C&D waste in order to meet the tight construction schedules, even if there are enough temporary rooms for C&D waste storage on site. In fact, there are often not enough places to pile up on-site residual C&D materials, and they are often treated as landfill waste as it is cheaper to do so.

Moreover, the problem associated with websites themselves is that there are too many websites with similar functionalities of waste exchange. Contractors can easily get confused to choose a suitable system. Moreover, the lack of user-friendly and efficient operational mechanisms often make current waste exchange websites unattractive. The authors also noticed that there was no waste exchange website that handles both Chinese and English in the user interfaces. Because of these, most waste-exchange websites could not attract enough users.

Nevertheless, the potential of waste exchange websites in disseminating information between contractors and buyers is well recognized. Because of the identified weaknesses of the TTS and the unattractiveness of existing waste exchange websites, the authors set to develop their own waste exchange website and integrate it with the TTS.

### 5.3.3 An e-commerce model

E-commerce has grown quickly in the construction industry as it is value-adding to business processes in the construction industry (DeMocker 1999; Berning and Diveley-Coyne 2000; NOIE 2001; Waugh and Makar 2001). According to the business model adopted, e-commerce systems can be categorized into three types: business-to-business model (e.g. e-IDC.com), business-to-customer model (e.g. Build.com), and combinatory model (e.g. EI-Internets.com). Because the business-to-business model has proven to be sustainable and profitable in the e-market of construction industry, it is most commonly used to develop E-commerce systems (Lais 1999), and more than 90 percent of architects, designers, and contractors expect to conduct more business over the Internet (Mark 2000). The authors thus select the business-to-business model to develop their online C&D waste-exchange system, which will be integrated with the TTS.

It should be mentioned that there is a waste-exchange website developed by the Environmental Protection Department of the HK Government in 2000 for C&D material management, from which practitioners in the Hong Kong construction industry can obtain useful information on waste minimization (WRC 2000). The developed website, named "C&D Material Exchange" (HKEPD 2002a,b,c), is open on the Internet. However, the system is incomplete when compared with other online waste-exchange websites such as those mentioned in Appendix C. For example, no list and search functions are provided in the system. In this regard, there is real potential to set up an online C&D waste exchange portal for the Hong Kong construction industry, and the authors attempted an e-commerce system as described below.

## 5.4 Integrated TTS-based e-commerce

### 5.4.1 Webfill model

Webfill is the e-commerce model for C&D waste exchange in Hong Kong developed by the authors, and it has been further developed to an online C&D waste exchange portal for the Hong Kong construction industry. Regarding the model design, different business models have been considered by the authors under the criteria to maximize recycle.

Rappa (2002) has summarized two essential strategic models for online exchange business including brokerage model and infomediary model. The brokerage model, e.g. Marketplace Exchange, provides a full range of services covering the transaction process, from market assessment to negotiation and fulfillment, for a particular industry. The exchange can operate independently of the industry, or it can be backed by an industry consortium. The broker typically charges the seller a transaction fee based on the value of the sale. There may also be membership fees. On the other hand, the infomediary model, e.g. Metamediary, facilitates transactions between buyers and sellers by providing comprehensive information and ancillary services, but does not get involved in the actual exchange of goods or services between the parties. Based on this theory, the infomediary model is finally selected for the integrated TTS-based e-commerce system.

Figure 5.3 illustrates the Webfill model with an information flowchart. The flowchart takes into account the common features of C&D waste exchange systems summarized in Figure 5.2 as well as the functional requirements of e-commerce.

Based on the background information mentioned earlier, the authors designed the Webfill model for five groups of main participants – construction contractors, property managers, manufacturers, recyclers, and landfill managers. The Webfill system is designed to provide member-oriented services such as add exchange information to the system, search for information for decision-making, and trade based on search results (trade options can be to sell waste or residual materials, to buy second-hand materials, to buy recovered or recycled materials), etc. An

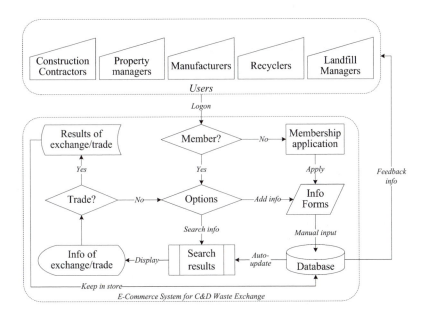

*Figure 5.3* Webfill e-commerce model for C&D waste exchange.

e-commerce server is designed to support the Webfill system, and all C&D waste information will be put into a database. E-mail service is used to link users and the Webfill system, and all membership information and trade information, etc. will be sent to each user from the database via e-mail. In addition, a credit card based online payment system is also adopted in the Webfill system to facilitate trade processes.

### 5.4.2  Users and their benefits

As mentioned earlier, there are five groups of potential users of the proposed Webfill system. Among them, construction contractors and property managers are providers of the C&D waste as well as the consumers of residual or second-hand materials, and recovered or recycled materials; manufacturers are providers of new materials made from raw materials or from C&D waste debris; landfill managers are providers of recyclable C&D waste debris, second-hand materials, and backfill materials, etc.; and recyclers are businessmen working among the construction contractors, the property managers, the manufacturers, and the land-fill managers to provide them information and transportation services. Table 5.3 describes the five kinds of users and their key roles together with benefits they can gain by using the Webfill system.

Benefits of using the Webfill system can be further elaborated as follows. First, as the TTS forces contractors to look for an inexpensive way to dispose of their C&D waste without paying tipping fees, the contractors can use the Webfill system in their best interests to find a buyer(s) for their residual construction

*Table 5.3* The usefulness of the Webfill system

| Users | Roles | User requirements | | Benefits |
|---|---|---|---|---|
| | | Sell | Buy | |
| Construction contractors | Waste generator | Recyclable waste Residual materials | Recovered materials Residual materials | Reduce tipping fee Reduce wastage Save on buying materials |
| Property managers | Waste generator | Recyclable waste Residual materials | Recovered materials Residual materials | Reduce tipping fee Reduce wastage Save on buying materials |
| Manufacturers | Material make Waste recovery | Recovered materials | Recyclable waste; Residual materials | Save on buying raw materials Increase sell |
| Recyclers | Waste trade | Recyclable waste Residual materials Recovered materials | Recyclable waste Residual materials Recovered materials | Increase waste re-use |
| Landfill managers | Waste disposal Waste trade | Recyclable waste | N/A | Decrease disposal of waste |

materials, who may be other contractors, manufactures, or recyclers. Contractors can also buy residual or used materials and equipment from other contractors, or buy inexpensive recovered materials from manufacturers, or deal with recyclers, in order to lower construction costs. Second, manufacturers can either sell their low-cost products made from recovered materials on the Webfill system at attractive prices to contractors, or buy cheaper raw and processed materials and used equipment from contractors. Recyclers and landfill managers can also sell their recovered products to contractors or manufacturers. Third, recyclers can either sell second-hand materials to contractors and manufacturers on the Webfill system, or buy cheap materials from contractors. Last but not least, landfill managers can either sell recyclable or recoverable materials to manufacturers and recyclers at low prices or free of charge on the Webfill system in order to reduce the total amount of C&D waste tipped at public filling facilities. Consequently, the Webfill system is able to attract construction contractors, property managers, manufacturers, recyclers, and landfill managers to work together as the Webfill system creates a win-win situation for all of them.

The Webfill system provides members a group of functions in e-commerce selections (refer to Appendix D). All selections are combined together according to the generic e-procurement process of Webfill that is described in Figure 5.3, and a demonstration website for local C&D waste exchange was located at http://158.132.107.159/mm/index.asp(2000/2003).

### 5.4.3   Website flexibility

#### 5.4.3.1   Membership

Users are required to register to become members of the Webfill system. After registration, Webfill provides every member with a trade account. A Webfill member can use the account ID and the self-determined password to login. Members enjoy a range of services including updated information on residual and reusable materials available, and functions for searching, ordering, selling, auctioning, and bidding of materials. The Webfill system automatically records the trading details of each member, which can further provide useful information for members to sell or buy materials. The trading records of a member are also used to assess his or her contribution to reducing C&D waste, and an annual reward system is used to encourage and reward active members.

#### 5.4.3.2   Commission fee

Incomes for the Webfill system can be generated in two ways: one is commission fee and another is advertisement fee. Webfill charges 0.1% of commission fee from each successful transaction. Every member is asked to provide the credit card information to prevent the evasion of commission fees. When the Webfill system sends email notification with a trade receipt to the seller and the buyer, as shown in Figure 5.3, the commission fee will be charged automatically from the seller's credit card account. If a buyer is not satisfied with what he ordered according to the Webfill's information and does not conclude the transaction with the seller, he can inform the Webfill system and the commission fee is then released.

## 5.5   Webfill simulation

Although a demonstration website of Webfill was developed, whether Webfill can really play the expected role in C&D waste reduction in Hong Kong is still a question. Besides research initiatives in a questionnaire survey form regarding the acceptance of the Webfill system, the Webfill model recalls a business process system, and the authors thus try to adopt the concept of e-commerce system simulation to experiment the Webfill system based on process simulation with statistical parameters relating to the generation, re-use, recycle and disposal of C&D waste in Hong Kong. The simulation which enables the authors to evaluate the performance of the Webfill system by comparing the results from two models, simple TTS and Webfill-enhanced TTS, is conducted. Considering the specific characteristics of the process flowcharts of the TTS and the Webfill system (refer to Figures 5.3–5.5), a commercial simulation software, i.e. ProcessModel (processmodel.com) is selected as the tool to simulate the simple TTS and the Webfill-enhanced TTS.

### 5.5.1   Simulation models

There are two basic steps involved in developing a simulation model: one is to establish a process model for simulation and another is to set some basic parameters according to real conditions. A process model is a process flow diagram that uses associated data to describe a real-life process, where objects (graphic shapes) and connections (lines connecting the graphic shapes) are used to represent process elements and relationships, respectively. In order to compare the simulation results between the TTS and the Webfill-enhanced TTS, two process-based simulation models are illustrated in Figures 5.4 and 5.5.

Figure 5.4 illustrates a simple TTS-based simulation model. There are ten entities in this model: *Materials* for both public buildings and private buildings, *Buildings* and *Civil Works* for the built environment, *Public Projects* for construction of public buildings, *Private Projects* for construction of private buildings, *On-site waste classification and storage* from both public and private projects, *Waste Recovery* at building material manufactories, the *TTS*, *Pre-landfill* for waste re-classification and storage, and *Landfill* for permanent waste disposals. All these entities are treated as processes inside the model, and the relations between any two entities are described using arrow lines with keyword indications.

Figure 5.5 illustrates the proposed Webfill-enhanced TTS simulation model. In addition to the ten entities inside the simple TTS-based simulation model,

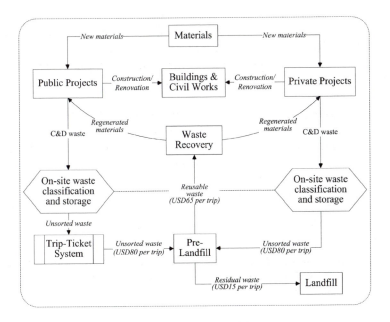

*Figure* 5.4 A simple TTS-based simulation model.

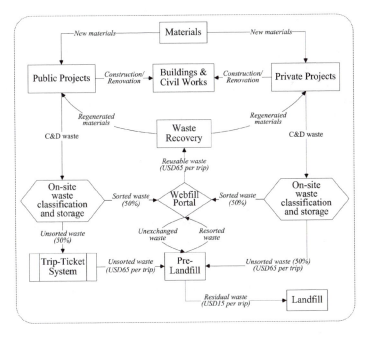

*Figure* 5.5 A proposed Webfill-enhanced TTS simulation model.

illustrated in Figure 5.4, a new entity of *Webfill* is integrated with the simple TTS-based simulation model. As the *Webfill* entity aims to introduce e-commerce into current TTS-based C&D waste management processes, this integration directly makes changes to the whole material chain in the Hong Kong construction industry. The influences include the following:

- for the two *On-site waste classification and storage* entities, Webfill divides part of C&D waste into e-commerce processes;
- for the *Waste Recovery* entity, Webfill interposes among the entities of *Waste Recovery*, *On-site waste classification and storage*, and *Pre-landfill* so as to provide more options to facilitate the recycle of C&D waste from construction sites and landfills; and
- for the *Pre-landfill* entity, Webfill provides a bridge to lead C&D waste disposed to the landfills back to the materials cycle.

Similar to Figure 5.4, all entities in Figure 5.5 are treated as processes inside the model, and relations between any two entities are described using arrow lines with keyword indications.

Regarding the five main participants involved in the two models – construction contractors, property managers, manufacturers, recyclers, and landfill managers – each of them occupies the relevant entity/entities inside the two models. For example, construction contractors and property managers have the same entities of *Public Projects*, *Private Projects*, and *On-site waste classification and storage*, etc; landfill managers have the entities of *Pre-landfill*, and *Landfill*; manufacturers have the entities of *Materials* and *Waste Recovery*. Although Figures 5.4 and 5.5 do not give entities to recyclers, it is generally regarded that recyclers can participate in the waste exchange at the entities of *Materials*, *Public Projects*, *Private Projects*, *On-site waste classification and storage*, *Waste Recovery*, *TTS*, and *Pre-landfill* to provide useful information on C&D waste recycle to other participants.

### 5.5.2 Basic parameters

Parameters have to be valued before running simulations based on the two models. In order to make a comparison between the two models as described in Figures 5.4 and 5.5, the authors decided to use the same set of parameters. Table 5.4 provides a list of some basic parameters selected for simulation, and Figures 5.4 and 5.5 provide necessary information for parameter settings. In addition to these parameter settings, the authors assume that the quantity of C&D waste ($G_{C\&Dwaste}$) generated by either the *Public Projects* or the *Private Projects* follows a normal probability distribution, and it is calculated using Equation 5.1.

$$G_{C\&Dwaste} = 0.036 \times N(20, 5) \tag{5.1}$$

Table 5.4 Parameters for the comparison simulation

| Adjusted items | System parameters | Real characteristics | Simulation settings |
|---|---|---|---|
| Process duration | 40 hour | 8 years (300 working days/year) | 1 min = 1 day |
| Waste quantity | 1 unit | 22,000 tons/day (Mean value of the statistic data from 1998 to 2003) | 1 unit = 11,000 tons waste rate is 3.6% $0.036 \times N(20,5)$ |

The assumption is made based on the details of the amount of C&D waste derived from the governmental statistic data (HKEPD 1999/2003) and relevant statistical analysis conducted by the authors.

### 5.5.3   Simulation results

Each simulation process sustains for about 25 minutes in Microsoft Windows XP operating system with Intel Pentium 1 GHz CPU and 512 MB RAM. Although the parameters are set based on historic data, it has been noticed that the simulation actually can provide more information regarding various business circumstances. However, as the purpose of this section is to provide a case study to demonstrate the process simulation that can be used to experiment an e-commerce system, the authors will not present more details about various experiments conducted on the two simulation models in accordance with various values of each parameter and further discuss regarding how to use feedbacks from simulation processes to revise a proposed business model. In this regard, the values of all parameters are kept in their original form as mentioned in the above context; and a group of simulation results and relevant comparisons are presented in Table 5.5.

Simulation analysis shows that the implementation of Webfill system in an 8-year period provides some foreseeable results. For example, the utility of land-fill decreases 85% and the TTS utility is reduced by 12%, while the utility of waste recovery increases 493%. These results indicate that the Webfill system can effectively reduce C&D waste disposed to the landfills and increase the use of recovered materials in building and civil works. Moreover, the total quan-tity of C&D waste is reduced by 8% on average between public projects and private projects; and the average waste cycle time increases 42%; the average value-added time of waste recovery lengthens 42%, and the average waste trans-portation cost increases 55%. These results indicate that the Webfill-enhanced TTS can reduce the amount of C&D waste at the landfill sites by increasing waste recovery and re-use.

The simulation reveals some unique results that other kinds of evaluation tools are unable to evaluate. This advantage is achieved by conducting process

*Table 5.5* Simulation results and comparisons

| Simulated Items | Simulation results | | Changing rates (%) |
| --- | --- | --- | --- |
| | TTS model | Webfill model | |
| Landfill utility (%) | 15.5 | 2.3 | −85 |
| Waste recovery utility (%) | 2.9 | 17.2 | +493 |
| TTS utility (%) | 5.1 | 4.5 | −12 |
| Quantity of C&D waste (unit) | | | |
| Public project | 328 | 302 | −8 |
| Private project | 336 | 309 | −8 |
| Average | 332 | 306 | −8 |
| Average waste cycle time (day) | | | |
| Public project | 7 | 10 | +43 |
| Private project | 5 | 7 | +40 |
| Average | 6 | 9 | +42 |
| Average value-added time (day) | | | |
| Public project | 3 | 4 | +33 |
| Private project | 2 | 3 | +50 |
| Average | 3 | 4 | +42 |
| Average waste transportation cost (USD) | | | |
| Public project | 90 | 140 | +56 |
| Private project | 85 | 130 | +53 |
| Average | 88 | 135 | +55 |

simulation in e-commerce system evaluation. Although it has been proved that the Webfill-enhanced TTS is more effective than simple TTS in C&D waste reduction, simulation results also indicate that the average waste transportation cost will increase, which means that the e-commerce system for C&D waste exchange will lead to more transportation from the construction industry, and more energy consumptions indeed.

## 5.6   Conclusions

This chapter presents a novel e-commerce simulation using a model e-commerce system, Webfill, which is integrated with the TTS used in Hong Kong for managing C&D waste disposal. The Webfill e-commerce system provides an on-line C&D waste exchange platform between construction contractors, property managers, construction material manufacturers and recyclers, and landfill managers. In order to evaluate the performance of the Webfill-enhanced TTS e-commerce system in reducing the C&D waste, a process-based simulation is done which allows the authors to directly compare the simple TTS and the Webfill-enhanced TTS. Simulation results indicate that the Webfill-enhanced TTS apparently reduces the total amount of C&D waste, through encouraging the increase of waste recovery. It is thus suggested that the Webfill-enhanced TTS be applied in the Hong Kong construction industry in order to deal with the continuously increasing amount of C&D waste. Furthermore, the Webfill simulation experiments a new area of e-commerce business plan evaluation, in which the concept of process simulation can be successfully implemented. Further research efforts should engage in Webfill model revision based on simulation results, and consider simulation parameters as well.

The successful application of process simulation in e-commerce business plan evaluation in this chapter reveals an emerging trend in e-commerce strategic management using quantitative approaches. Because process simulation is generally accepted in business management, it is an economical way to directly use commercial process simulation package for e-commerce simulation. However, as there are some limitations in process simulation packages such as no permission for users to modify internal and external business environmental factors based on their various experiments, it is essential to use current business strategic management tools such as the SWOT analysis in e-commerce system evaluation as complements. In this regard, further research tasks are required to integrate current qualitative strategic management tools into business process simulation environment.

# Chapter 6

# Knowledge-driven evaluation

## 6.1 Introduction

The adverse environmental impacts of construction such as soil and ground contamination, water pollution, construction and demolition waste, noise and vibration, dust, hazardous emissions and odours, demolition of wildlife and natural features, and archaeological destruction have been a matter of concern since the early 1970s and are of more and more academic and professional interest in the construction industry especially after the ISO 14000 series of EM standards was enacted. In this regard, quantitative analytical approaches to EM in construction are currently not as prevalent as qualitative approaches, such as regulations and practical guides, due to the difficulties in transformation of practical data to abstract data that are necessarily used in calculation for EM. However, it is hard to accept an EMS without the background support of quantitative analytical approaches, or an EMS is not consummate if adequate quantitative analytical approaches for sustainment are not there. For the sake of practical approaches and their integrated application for quantitative EM in construction, a research project, *An Integrated Analytical Approach to Environmental Management in Construction* (Chen 2003), was set up in the Research Centre for Construction Innovation, the former Research Centre of Construction Management and Construction IT, Department of Building and Real Estate, the Hong Kong Polytechnic University in 1999 and the findings from this research project include one holistic approach and four quantitative EM tools for environmental-conscious construction project management.

The successful implementation of ISO 14001 EMS in construction sectors requires far more than just the apparent prevention and reduction of negative environmental impacts in a new project development cycle as well as each proposed construction process cycle during pre-construction stage, continuous improvement of the environmental management function based on institutionalization of change throughout an enterprise to reduce pollution during construction stage, and efficient synergisms of pollution prevention and reduction such as waste recycle and regeneration in the construction industry throughout construction stage and post-construction stage. It necessitates a complete reengineering and transformation of all organizational functions related to the project-based construction management (CM) throughout each construction stage in environment-conscious construction sectors. In addition to the integration of all stages of the construction life cycle, the effective implementation of ISO 14001 EMS in construction enterprises also demands functional tools to facilitate the deployment of the EMS throughout construction enterprises and construction projects in both macro and micro environment's for organizational sustainable development. The lack of effective EM tools and the insufficient utilization or abuse of EM tools can directly obstruct the implementation of EMS in either construction enterprises or the construction industry even though such a management system has been accredited individually in advance (Chen and Li *et al.* 2002a). For example, according to a recent statistical data analysis conducted in the Chinese construction industry (Chen and Li *et al.* 2004a), the annual rate of environmental impact assessment (EIA) approvals for new construction projects was 97% in 2001, whilst the rate of the ISO 14001 EMS accreditations for construction enterprises was as low as 1‰ in mainland China. It is obvious that approval rate of EIA is much higher than the accreditation rate of EMS in this case; however, it also discloses that most construction enterprises have not yet adopted or accepted the ISO 14001 EMS in mainland China. As the EIA approval is only required at registration stage of construction projects and the EMS implementation is normally required to be sustained during the whole period of construction projects, the disagreement existing between the two rates discloses that there may be little coordination between the EIA process and EMS implementation in construction projects, and thus the EIA may not really serve as a tool to promote EM in construction projects in the construction industry in mainland China. Although the authors have not collected enough data to support the statement that the implementation rate of mandated EIA process is universally higher than the accreditation rate of encouraged ISO 14001 EMS in the construction industry all over the world, except for the Chinese case mentioned above, some indirect evidences can be presented based on previous research reports related to the implementation of ISO 14001 EMS in construction sectors in different countries such as Australia (CPSC 2001; Zutshi and Sohal 2004a,b), China (Chen, Li and Wong 2000; Lo 2001; Tse 2001; Zeng *et al.* 2003), Singapore (Kein *et al.* 1999; Ofori *et al.* 2000), UK (CIRIA 1999), and USA (Darnall 2001; Valdez and Chini 2002), in which construction sectors emphasized that the procedure of the EIA or the EMS should be undoubtedly adopted under mandatory instructions from local governments and EM tools were specially required to facilitate the implementation of the EIA and the EMS in project-based construction management.

Regarding the EM tools for construction sectors, as quantitative EM tools are currently not as regularly adopted as qualitative EM tools such as administrative regulations and practical guides due to the difficulties in raw on-site information and data transformation for necessary computation in the EM-integrated construction project management, it is necessary to power an EMS accredited or under accreditation with adequate support from quantitative EM tools and their background knowledge warehouse, which is the essential component of an enterprise's Knowledge Management System (KMS), where knowledge is developed, stored, organized, processed, and disseminated (SAP INFO 2004). Based on this consideration, the authors want to put forward a novel methodology entitled E+ in which a KMS for environmental-conscious construction project management is integrated with EM tools and dynamic EIA process transplanted from a combination of a standard EMS process and a static EIA process. It is expected that the deployment of E+, or the knowledge-driven EMS-based dynamic EIA process, can facilitate KM initiatives for improved competitiveness of construction enterprises in EM. This holistic objective will be achieved step by step through the following four sub-objectives:

1   to illustrate an integrative knowledge-driven EM prototype to capture and re-use data, information, and knowledge for dynamic EM in construction project management;
2   to describe quantitative EM tools which can be integrated into the integrative KM model for dynamic environmental-conscious construction project management;
3   to describe an interaction of quantitative EM tools with the integrative KM model and key information techniques to for a KMS to support dynamic EM in construction project management; and
4   to demonstrate the implementation of the KMS through a case study.

First of all, an integrative methodology for dynamic EM in construction project management is developed as a comprehensive frame prototype entitled E+. Next, four quantitative approaches to be integrated into the E+ model are developed step by step. They are the analytical approaches for construction planning such as the CPI method and evaluation of environmental-conscious plan alternatives named env.Plan method, and analytical approaches for construction logistics management such as the IRP method, and construction waste exchange model named Webfill method. After that, two knowledge management entities – knowledge capture entity and knowledge re-use entity – together with six kinds of relative CM knowledge bases are unified into the E+ model aimed for integrative effectiveness and efficiency of the model. Finally, the implementation of the integrative analytical approach is demonstrated with an experimental case study.

For the integrative methodology of KMS for EM in construction, this chapter mainly contributes to existing theory or EM in construction in the area of quantitative analytical approaches and their integrative implementation. According to the literature review and questionnaire survey for this research, the lack of effective, efficient, and economical (E3) quantitative analytical approach is one of the obstacles to implementing EM in construction. Therefore, there are four points of contributions from this research to the existing theory or practice of EM in construction:

1   This research has developed an integrative methodology (E+) to implementing EMS and KM in construction, with a rigorous dynamic EIA model based on various functionally different approaches to EM in a construction

cycle. The E+ prototype was originally created in both the theory and practice for EM in construction, and it is open to further integration of various functionally different approaches for EM in construction other than the three EM tools presented in this chapter. Because the E+ is both EMS-oriented and process-oriented in construction, it can help contractors to implement EM from a messy situation to a normalized system and to effectively share EM knowledge and information internally and externally.

2   The CPI method integrated in the E+ model is a quantitative approach to predicting and levelling complex adverse environmental impacts potentially generated from construction and transportation due to the implementation of a construction plan. As a result, the CPI method has been integrated into E+ EM Toolkit A, one functional section of E+ system, to carry out the task in environmental-conscious construction planning.

3   The IRP method is a quantitative approach to reducing wastage of construction materials on a construction site, and it is designed to be effectively implemented by using a bar-code system. The IRP is then integrated in the E+ EM Toolkit B, another functional section of E+, as a basic component.

4   The Webfill method is an E-commerce model designed for the trip-ticket system to effectively reduce, re-use, and recycle C&D waste. Although there is lack of data to prove the efficiency in reality, the computer simulation results and a questionnaire survey from another research (Chen 2003) have proved that the Webfill system can effectively realize the design function. As a result, the Webfill is also integrated in the E+ model as an important component of the E+ EM Toolkit C.

The authors expect that readers can obtain state-of-the-art socio-technical perspectives from the introduction of the E+ prototype and its Toolkits, and know how E+ can work for a dynamic EIA process in construction with integrated supports from E3 quantitative analytical approaches in the Toolkits.

## 6.2  Background

EM in construction has received more and more attention since the early 1970s. For example, studies on noise pollution (U.S.EPA 1971), air pollution (Jones 1973), and solid waste pollution (Skoyles and Hussey 1974; Spivey 1974a,b) from construction sites were individually conducted in the early 1970s. Although the expression of EM in construction came out in the early 1970s after the U.S. National Environmental Policy Act of 1969 enacted (Warren 1973), the concept of EM in construction was introduced in the late 1970s, when the role of environmental inspector was defined in the design and construction phases of projects to provide advice to construction engineers on all matters in EM (Spivey 1974a,b; Henningson 1978). However, there had been little enthusiasm for establishing an EMS in construction organizations until two main important standards, BS 7750 (released by the BSI Group in 1992) and the ISO 14000 series (released by the ISO in 1996), were promulgated to guide the construction industry from passive CM on pollution reduction to active EMS for pollution prevention.

In the 1990s, the CIRIA conducted a series of reviews on environmental issues and have undertaken initiatives relevant to the construction industry after the introduction of BS 7750 (Shorrock *et al.* 1993; CIRIA 1993, 1994a,b, 1995; Guthrie and Mallett 1995; Petts 1996). Thereafter, research works on EM have also focused on the implementation of EMS and the registration of ISO 14001 EMS by authoritative institutions in the construction industry, such as the CIOB (Clough and Antonio 1996), the FIDIC (1998), the Construction Policy Steering Committee (CPSC 1998), and the CIRIA (Uren and Griffiths 2000).

In order to assess the extent of EMS implementation within the construction industry, several investigations have been conducted independently by researchers in different countries since 1999. For example, Kein *et al.* (1999) conducted a field study in Singapore to assess the level of commitment of ISO 9000-certified construction enterprises to EM. They found that contractors in Singapore were aware of the merits of EM, but were not instituting systems towards achieving it; Ofori *et al.* (2000), also in Singapore, then conducted a survey to ascertain the perceptions of construction enterprises on the impact of the implementation of the ISO 14000 series on their operations. Major problems were identified, such as the shortage of qualified personnel, lack of knowledge of the ISO 14000 series, indistinct cost–benefit ratio, disruption and high expenses on changing traditional practices, and resistance from employees, etc.; the CIRIA (1999) led a self-completion questionnaire survey of the state of environmental initiatives within the construction industry and of sustainability indicators for the civil engineering industry in the United Kingdom; Tse (2001) conducted an independent questionnaire survey in the Hong Kong construction industry to gain a further understanding of the difficulties in implementing the ISO 14000 series; Lo (2001), also in Hong Kong, made an effort to identify nine critical factors for the implementation of ISO 14001 EMS in the construction industry based on critical factors drawn from an investigation in another industry; the CPSC (2001), in Australia, conducted a questionnaire survey on the New South Wales construction industry on EM with industry leaders; Chen and Li *et al.* (2004b) conducted a questionnaire survey of main contractors in five main cities in mainland China and found that there are five classes of factors influencing the acceptability of the ISO 14000 series of EM standards – governmental regulations, technology conditions, competitive pressures, cooperative attitude, and cost–benefit efficiency; besides this, Zeng *et al.* (2003) also conducted a questionnaire survey on the mainland China construction industry to discover the conditions of implementations of the ISO 14000 series. All these questionnaire surveys aimed to clarify the real situations in the adoption and implementation of the ISO 14000 series of EM standards in the local construction industries, and provided relative perspectives on how to conduct EM to the construction industry.

One important contribution of these surveys is that researchers have obtained useful insights into the problems and difficulties of implementing the ISO 14000 series in construction. Their survey results provide useful information not only for improving efficiency of EMS implementation but also for developing the EMS itself, focusing on highly effective and economical EM in the construction industry. For example, Tse (2001) has found four major obstacles in implementing the ISO 14000 series in the Hong Kong construction industry – lack of government pressure, lack of client requirement or supports, expensive implementation cost, and difficulties in managing the EMS with the current sub-contracting system. One cannot easily draw such constructive conclusions in detail without such a kind of survey.

Besides these questionnaires used to survey the implementation of the ISO 14000 series in the construction industry in different countries, case studies are further applied to investigate the acceptability of the ISO 14000 series to the construction industry. For example, Valdez and Chini (2002) conducted a literature search and a case study of a construction contracting firm certified for the ISO 14001 EMS in the United States. They concluded that the positive aspects of certification outweigh the negative aspects and recommended to add government support and the combined use of the ISO 14000 series with other EM methods and matrices.

On the other hand, the remarkable difference between the rate of ISO 14001 EMS accreditation and EIA implementation in some countries indicates that contractors there have not really implemented EM and accepted the ISO 14000 series (Chen 2003). The EIA of construction projects is a process of identifying, predicting, evaluating, and mitigating the biophysical, social, and other relevant environmental effects of development proposals or projects prior to major decisions being taken and commitments made (IAIA 1997). Although the EIA has been accepted by the construction industry in different countries according to governmental regulations to evaluate the environmental impacts of a construction project, the implementation rate of ISO 14001 EMS accreditation in the construction industry is normally much lower than the implementation rate of EIA. For example, according to the *Official Report on the State of the Environment in China 2001* (China EPB 2002), the annual implementation rate of EIA for construction projects was 97% in 2001 in mainland China. In addition, a further investigation on the implementation rate of EIA in mainland China indicates that the average EIA rate from 1995 to 2001 was 88%, with an increasing rate of 23% (China EPB 2002). By contrast, the percentage of construction enterprises that have been awarded environmental certificates versus total government-registered construction enterprises in mainland China is as low as 0.083% (Chen 2003). Statistical figures also indicate that most construction enterprises have not yet adopted or accepted the ISO 14000 series in mainland China. Because of the disagreement between the implementation rates of EIA and EMS, there may be little coordination between the EIA process and EMS implementation in construction projects, and thus the EIA may not really serve as a tool to promote EM in the construction industry in those countries. For that reason, adverse environmental impacts such as noise, dust, waste, and hazardous emissions still occur frequently in construction projects in spite of their EIA approvals prior to construction.

Besides the status of implementing EIA in construction, the authors also noticed the emerging willingness to apply KM in the construction industry. There is growing consciousness, requirements, and initiatives of KM in order to manage the intellectual capital and get benefits from previous construction processes and projects (Zyngier 2002; Zarli *et al.* 2003). For example, the C-Sand project (c-sand.org.uk) has been conducted in the UK to foster organizational practices which enable knowledge creation for subsequent sharing and re-use, and to promote sustainable construction (Khalfan *et al.* 2003). As one of the largest contracting companies in the United States, Centex Construction Group (centex-construction.com) faces some knowledge-related business challenges that are not always associated with the construction industry. For instance, they have a technology infrastructure in place where all professionals in the company have computing power, i.e. laptops and/or desktops. All offices and job sites are connected to a nationwide WAN via dial, ISDN, and Frame. Remote access is Web-based and available from anywhere to lead some initiatives to increase knowledge sharing and provide better information access across the company's diverse landscape (Velker 1999). Beyond the development of knowledge warehouse (KW) in the construction industry, socio-technical research also reflected that a majority agreed to the statement that KM is an extension of IT (Zyngier 2002). The progress of KM in construction also reflects the trend of construction enterprises away from traditional blue-collar operations towards a more knowledge-based CM.

According to the survey results, the implementation of the EMS requires EM-support approaches as practicable as the EIA approach, which is popular and easier to use by contractors. That is, although the governmental regulations have been identified as a major factor influencing the implementation of the EMS and the EIA in the construction industry according to surveys and case studies mentioned above, the construction industry is still a negative receiver if there are not enough technology conditions to support the implementation of EMS, especially the techniques or tools which can help contractors to conduct EM in construction projects where the most amount of negative environmental impacts are generated. Even for the positive bodies in the construction industry that have high willingness to implement EM, effective, efficient, and economical EM tools are essential. Moreover, the requirements of re-use EM experience also exist (Chen 2003). Based on this consideration, the authors will integrate several EM tools and an EMS-based dynamic EIA process developed previously into an environment of KM entitled E+ for effective, efficient, and economical EM in dynamic construction project management.

## 6.3   The E+

### 6.3.1   Methodology

The E+ is an integrative methodology for effective, efficient, and economical EM in construction projects in which an EMS-based dynamic EIA process is applied within a knowledge support system for active knowledge capture and re-use about environmental-conscious CM during construction. The successful implementation of an EMS in construction projects requires far more than just the apparent prevention and reduction of adverse or negative environmental impacts in a new project and its construction process development cycles during pre-construction stage, continuous improvement of the EM function based on institutionalization of change throughout an on-site organization to reduce pollution during construction stage, or efficient synergisms of pollution prevention and reduction such as waste recycle and regeneration during construction and post-construction stages. It necessitates a complete transformation of CM in an environmental-conscious enterprise, such as changes in management philosophy and leadership style, creation of an adaptive organizational structure, adoption of a more progressive organizational culture, revitalization of the relationship between the organization and its customers, and rejuvenation of other organizational functions (such as human resources engineering, research and development, finance, marketing, etc.) (Azani 1999). In addition to the transformation for the EM in construction enterprises, the integrative methodology, E+, for the effective, efficient, and economical implementation of the EM in all phases of the construction cycle including the pre-construction stage, the construction stage and the post-construction stage is necessarily activated, together with other rejuvenated CM functions such as human resources, expert knowledge, and synergetic effect.

There are already some approaches to effectively implementing the EM on site at different construction stages. For example, the CPI approach which is a method to quantatively measure the amount of pollution and hazards generated by a construction process and construction project during construction by indicating the potential level of accumulated pollution and hazards generated from a construction site at the pre-construction stage can be utilized (Chen, Li and Wong *et al.* 2000), and by reducing or mitigating pollution levels during the construction planning stage (Li *et al.* 2002); in addition to the CPI approach, an ANP approach to construction plan selection (Chen and Li *et al.* 2003a,b,c,d), a life-cycle assessment (LCA) approach to material selection (Lippiatt 1999), and a decision programming language (DPL) approach to environmental liability estimation (Jeljeli and Russell 1995) also provide quantitative methods for making decision's on EM at pre-construction stage; for the construction stage, a crew-based IRP approach, which is realized by using bar-code system, can be utilized as an on-site material management system to control and reduce construction waste (Chen and Li *et al.* 2002a); for the post-construction stage, an online waste exchange (Webfill) approach which is further developed into an e-commerce system based on the trip-ticket system for waste disposal in Hong Kong can be utilized to reduce the final amount of C&D waste to be landfilled (disposed of the C&D waste in a landfill) (Chen and Li *et al.* 2002b). Although these approaches to EM in construction projects have proved to be effectively, efficiently, and economically applicable in a corresponding construction stage, it has also been noticed that these EM tools can be further integrated for a total EM purpose in construction projects based on the interrelationships among them. The integration can bring about not only a definite utilization of current EM tools but also an improved environment for contractors to maximize the advantages of utilizing current EM tools due to sharing EM-related data, information, and knowledge in construction project management.

As mentioned earlier, the EMS is not as acceptable as the EIA at present in some countries such as mainland China partly due to the lack of effective, efficient, and economical EM approaches in construction besides the governmental regulations, and the tendency of practical EM in construction is to adopt and implement the EMS when the EIA report/form of a construction project has been approved. As a result, the E+ for contractors to enhance their environmental performance, which integrates all necessary EM tools available currently, is just appropriate.

The proposed E+ aims to provide high levels of insight and understanding regarding the EM issues related to the project management in a construction cycle. In fact, the current EIA process applied in construction projects is mainly conducted prior to the pre-construction stage, when a contractor is required to submit an EIA report/form based on the size and significance of the project and the EIA process for the construction stage is seldom conducted in normalized forms. Due to the strong alterability of the environmental impacts in the construction cycle, commonly encountered static EIA processes prior to construction cannot accommodate the implementation of the EMS in project construction, and a dynamic EIA process is thus designed for the E+. In addition, current EM tools are to be combined with a frame of ISO 14001-required EMS process (a process of the EMS including issuing environmental policies, planning, implementation and operation, checking and corrective action, and management review) (refer to Figure 2.2) according to their interrelationships, with which various EM-related data, information, and knowledge in construction can be captured, organized, and re-used. Because the main task of EM in the construction cycle is to reduce adverse environmental impacts, the dynamic data transference in the framework is the prime focus of the E+ methodology. Thus, a prototype of the E+ is further put forward (refer to Figure 6.1).

Comparing with the original E+ model that was earlier developed in the plain integration of two EM tool entities – E+ Plan entity for environmental-conscious

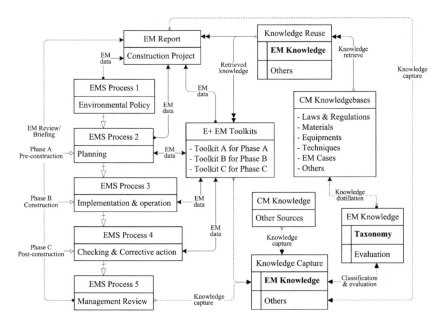

*Figure 6.1* The prototype of the E+ model.

planning at pre-construction stage and E+ Logistics entity for implementing EM at both construction stage and post-construction stage with the standard EMS process (Chen *et al.* 2004a) – the E+ prototype being discussed overcomes the limitations in reusing CEM knowledge that exist in the former E+ model by means of an embedded knowledge-driven procedure, and the conceptive E+ prototype is described in Figure 6.1.

The framework of E+ prototype comprises three main sections, including an E+ EM Toolkits entity, an E+ KMS entity, and an EMS-based EIA entity (see Figure 6.1). Features of each entity are described below:

1   The E+ EM Toolkits entity is the core of the E+ prototype, which consists of three kinds of EM tools corresponding to the three phases of a construction cycle – Toolkit A for pre-construction stage, Toolkit B for construction stage, and Toolkit C for post-construction stage.

2   The E+ KMS entity is the knowledge engine of the E+ prototype, which consists of five KM-related subentities: Knowledge Source, Knowledge Capture, Knowledge Classification and Evaluation, Knowledge Storage, and Knowledge Re-use subentities.

3   The EMS-based EIA entity is the essential structural frame of the E+ prototype, which consists of six EMS-related subentities: Environmental Policy, EM Planning, EM Implementation and Operation, EM Assessment,

EM Review, and EM Report subentities. These subentities belong to a standard EMS process normalized by the ISO 14000 series of EM standards.

### 6.3.2 Implementation

The implementation of the E+ prototype in CM needs an integrative software environment in which various E+ entities – the E+ EM Toolkits entity, the E+ KMS entity and the EMS-based EIA entity – can work together with the EMS process to accommodate both intramural and extramural EM-related assessments. Considering the general process of KM comprising knowledge planning, creation, integration, organization, transference, maintenance, and assessment (Rollett 2003) and the general process of computer software development, the authors decided to realize an E+ system through three main steps, as described below:

1  First step: feasibility study. The feasibility study is conducted not only prior to the establishment of the E+ model, but also before system analysis and development of the E+ software environment. First of all, it is important to analyse whether such an E+ system is necessary for the EMS-based dynamic EIA process in project construction, and this is to be done prior to the establishment of the E+ model. Next, if the E+ system is necessary, it requires a search for enough quantitative EM tools to support the E+ system, and this is to be done before system analysis and development of the E+ system. The feasibility study is essential for both a practicable E+ prototype and an effective, efficient, and economical E+ system.

2  Second step: system analysis and realization. System analysis and realization are to be conducted after the E+ model has been established. The aim of this step is to transform the E+ system from a model to a software environment with computer programming. Constrained by the length of this chapter, no further discussion is presented here to illustrate the development of the E+ system.

3  Third step: system evaluation. System evaluation is a trial process for the developed E+ system. There is also no further discussion related to this step as the E+ system is under construction. However, an experimental case study is conducted below to demonstrate the effective, efficient, and economical EM function of the E+ system.

The following discussions focus on several core EM tools adopted in the E+ model, and interrelationships among these EM tools while working for the EMS-based dynamic EIA process. As the EM tools selected for the E+ model in this chapter are the CPI approach to indicate adverse environmental impacts at pre-construction stage, the IRP approach for material management on site at construction stage, and Webfill approach for residual and waste material

and equipment exchange at post-construction stage, no more EM tools are discussed here.

## 6.4 EM tools for the E+

### 6.4.1 CPI

In the prototype of the E+, the CPI approach is set to the E+ EM Toolkit A
for construction planning at pre-construction stage (refer to Figure 6.1). The E+
EM Toolkit A captures data from three kinds of sources:

1   Source one: EMS Process, including EMS Processes 2 and 3.
2   Source two: E+ EM Toolkits, including Toolkits B and C.
3   Source three: knowledge bases (KBs) of EM, including knowledge base
    of environmental law and regulation, environmental-friendly construction
    materials, environmental-friendly construction machines, environmental-
    friendly construction techniques, EM cases, etc.

Meanwhile, the E+ EM Toolkit A transfers data to these three kinds of data
sources, and to the EM report such as the dynamic EIA report of a construction
project.

A quantitative approach named construction pollution index (CPI) (refer to
Chapter 3) is adapted in the E+ EM Toolkit A to evaluate adverse environmental impacts in construction planning at pre-construction stage. The CPI is
an approach to quantitatively measure the amount of pollution and hazards that
will be generated by a construction project or a construction process during
construction. The method measures CPI as shown in Equation 6.1.

$$\text{CPI} = \sum_{i=1}^{n} \text{CPI}_i = \sum_{i=1}^{n} h_i \times D_i \tag{6.1}$$

where CPI is Construction pollution index of an urban construction project; $\text{CPI}_i$
is Construction pollution index of a specific construction operation $i$; $h_i$ is hazard
magnitude per unit of time generated by a specific construction operation $i$; $D_i$
is duration of the construction operation $i$ that generates hazard $h_i$; and $n$ is
number of construction operations that generate pollution and hazards.

In Equation 6.1, parameter $h_i$ is a relative value indicating the magnitude of
hazard generated by a particular construction operation in a unit of time. Its
value is normalized into the range [0, 1]. If $h_i = 1$, it means that the hazard
can cause fatal damage or be catastrophic to people and/or properties nearby.
For example, if a construction operation generates noise and the sound level at
the receiving end exceeds the "threshold of pain", which is 140 dB (McMullan
1998) then the value of $h_i$ for this particular construction operation is 1. If $h_i = 0$,
then it indicates that no pollution and hazard is detectable from a construction

operation. It is possible to identify values of $h_i$ for all types of pollution and hazards generated by commonly used construction operations and methods based on users' experience and expert opinions.

Because the value of CPI reflects the accumulated amount of adverse environmental impacts generated by a construction project within its project duration, its utilization in construction planning is easily realized though a CPI histogram, similar to the resource histogram in a Gantt chart which is used in construction scheduling. By integrating the concept of CPI into a commonly used tool for construction project management such as Microsoft Project©, a system to neatly combine EM with project management is then formed, and project managers can use the CPI histogram to identify the periods in which the project will generate the highest amount of pollution and hazards, and reschedule the whole project to level extremely high CPI (refer to Chapter 3).

However, with respect to further reusing CM knowledge to define the $CPI_i$ (CPI of a specific construction operation $i$) for each process in different construction projects, the authors noticed that experts' opinions varied from project to project regarding the value of $CPI_i$. This means that the topic of reusable CM knowledge to define the current experience-based $CPI_i$ in construction planning has aroused discussion, and therefore the development of a new tool to suit the computation of CPI to facilitate the re-use of experts' knowledge at pre-construction stage is required. The tool for CM knowledge re-use to define $CPI_i$ is developed by using artificial neural network (ANN) approach. As the ANN-based approach to define $CPI_i$ has already been discussed in previous works (Chen and Li *et al.* 2004a,b), it is just adopted in the E+ prototype as an EM tool in the E+ EM Toolkit A.

### 6.4.2  IRP

In the prototype of the E+, the IRP approach (refer to Chapter 4) is set to the E+ EM Toolkit B for construction at construction stage (refer to Figure 6.1). The E+ EM Toolkit B captures data from three kinds of sources:

1  Source one: EMS Process 3.
2  Source two: E+ EM Toolkits, including Toolkits A and C.
3  Source three: KBs of EM, including knowledge base of environmental law and regulation, environmental-friendly construction materials, environmental-friendly construction machines, environmental-friendly construction techniques, and EM cases.

Meanwhile, the E+ EM Toolkit B transfers data to these three kinds of data sources, and to the EM report such as EIA report of a construction project.

The IRP approach (refer to Chapter 4) is an approach to quantatively measure the amount of material waste generated by a construction project or a process during construction. IRP measures the exact amount of material saved or wasted by each crew during construction, as shown in Equation 6.2.

$$C^i(j) = \sum_n \Delta Q^i(j) \times P_i = \sum_n \left\{ Q^i_{es}(j) - \left[ Q^i_{de}(j) - Q^i_{re}(j) \right] \right\} \times P_i \qquad (6.2)$$

where $C^i(j)$ is the total amount of material $i$ saved (if $C^i(j)$ is positive) or wasted (if $C^i(j)$ is negative) by crew $j$; $\Delta Q^i(j)$ is the extra amount of material $i$ saved (if the amount is a positive value) or wasted (if the amount is a negative value) by crew $j$; $Q^i_{es}(j)$ is the estimated quantity that includes the statistic amount of normal wastage; $Q^i_{de}(j)$ is the total quantity of material $i$ requested by crew $j$; $Q^i_{re}(j)$ is the quantity of unused construction materials returned to the storage by crew $j$; $P_i$ is the unit price for material $i$; and $n$ is the total number of tasks in the project that need to use material $i$.

According to the Equation 6.2, for a particular type of material $i$, the performance of crew $j$ in terms of material wastage can be measured by $\Delta Q^i(j)$, and at the end of the project, the overall performance of crew $j$ can be rewarded in agreement with $C^i(j)$. That is, the IRP is implemented according to the amount of materials saved or wasted by a crew, i.e. if a crew saves materials ($\Delta Q^i(j) > 0$), the crew will be rewarded based on the quantity of $C^i(j)$.

As the computation of IRP is done by measuring the exact amount of material saved or wasted during the construction process and comparing it with the estimated quantity of materials that will probably be consumed based on the statistic amount of normal wastage in other construction processes (see Equation 6.2), there is also a space to re-use CM knowledge to define the value of $Q_i(j)_{es}$. As the adoption of the ANN approach in quantity survey (Adeli and Karim 2001) for CM retrieval and re-use has received wide recognition in construction, the authors of further employed another ANN model (Chen *et al.* 2004a,b) to support knowledge re-use in IRP computation in the E+ prototype.

### 6.4.3 Webfill

In the prototype of the E+, the Webfill approach (refer to Chapter 5) is set to the E+ EM Toolkit C for post-construction work at post-construction stage (refer to Figure 6.1). The E+ EM Toolkit C captures data from three kinds of sources:

1 Source one: EMS Process 4.
2 Source two: E+ EM Toolkit B.
3 Source three: KBs of EM, including knowledge base of environmental laws and regulations, environmental-friendly construction materials, environmental-friendly construction machines, environmental-friendly construction techniques, and EM cases.

Meanwhile, the E+ EM Toolkit C transfers data to these three kinds of data sources, and to the EM report such as the dynamic EIA report of a construction project.

The Webfill approach (refer to Chapter 5) is an e-commerce method to increase the amount of C&D waste exchanged for re-use and recycle among different construction sites and material-regeneration manufactories. Disposal of C&D waste to landfills is usually charged in many countries (Chen 2003). For example, in order to orderly dispose C&D waste to disposal facilities by trucks, the TTS was implemented in the Hong Kong construction industry in 1999, which requires contractors to pay for the disposal of their C&D waste in terms of waste disposal receipts issued to them. The Webfill approach sets the TTS-based e-commerce model conforming to the external requirement, and simulation results indicate that the Webfill-enhanced TTS can apparently reduce the total amount of C&D waste through encouraging the increase of waste re-use and recycle.

The trade promotions of the Webfill system include an annual reward system and a finite release of commission fee based on the trading records of each member. Two kinds of EM-related data, which the Webfill system provides based on its trading records, can be used to indicate the environmental-conscious performance of contractors. They are the quantity of C&D waste a contractor sold $(Q_{sold})$ and the quantity of regenerated materials or reusable material a contractor bought $(Q_{bought})$. By using these two kinds of EM-related data, Equation 6.2 can be further developed into Equation 6.3. According to Equation 6.3, if the waste generated by crew $j$ is sold through the Webfill system or the crew $j$ request regenerated material bought from the Webfill system, the crew can thus be rewarded.

$$C^i(j) = \sum_n \Delta Q^i(j) \times P_i = \sum_n \{[Q^i_{es}(j) - (Q^i_{de}(j) - Q^i_{re}(j))]$$
$$+ Q^i_{sold}(j) + Q^i_{bought}(j)\} \times P_i \qquad (6.3)$$

where $Q^i_{sold}(j)$ is the quantity of waste material $i$ sold by or related with crew $j$; and $Q^i_{bought}(j)$ is the quantity of regenerated material $i$ requested by crew $j$.

The Webfill in the E+ EM Toolkit C plays a supporting role in CM knowledge retrieval and re-use by providing statistic data to define the value of $Q_i(j)_{es}$ required in both Equations 6.2 and 6.3. All statistic data from Webfill can be further used for the ANN model too.

### 6.4.4  Interrelationships

The interrelationships among the EMS process, the EIA process, the EM Toolkit, and the *Knowledge Capture* process and *Knowledge Re-use* process can be put up in agreement with EM-related data transferences. There are six kinds of EM-related data transferences in the E+ system. The first kind of data transference occurs between the EMS process and the EIA process; the second kind of data transference occurs among the EM Toolkits and the EIA process; the third kind of data transference occurs among the various EM Toolkits; the fourth kind of data transference occurs from the Knowledge Re-use entity to the EM

Toolkits and the EIA process; the fifth kind of data transference occurs from the EM Toolkits, the EIA process and the EMS process to the *Knowledge Capture* entity; and the sixth kind of data transference occurs from *Knowledge Capture* entity to the *Knowledge Re-use* entity through several essential KBs such as KB of environmental law and regulation, KB of environmental-friendly construction materials, KB of environmental-friendly construction machines, KB of environmental-friendly construction techniques, and KB of EM cases, etc. Because all these data are generated from different construction stages, integrative data transference in the E+ system can thus provide up-to-date information to the EIA process and the dynamic EIA process is realized accordingly. In order to completely clarify the interrelationships potentially existing in the E+ system, some of the EM-related data and their transferences are summarized in Table 6.1.

For the knowledge-driven E+ system, a proper way of representing EM-related knowledge is essential to influence its effectiveness. Knowledge representation, as one of the central and in some ways most familiar concepts in artificial intelligence, is best understood in terms of the five fundamental roles that it

*Table 6.1* Interrelationships among EM-related data in the E+ system

| Data host | Data name | Transfer to | Received from | Usefulness |
|---|---|---|---|---|
| CPI host (Toolkit A) | CPI $CPI_i$ | EIA host, KBs | – | Data update for EIA report |
| | $CPI_i$ | EMS process②, KBs | – | Construction planning |
| | $CPI_{waste}$ | IRP host | – | Quantity survey of waste |
| | $h_i$ | KBs | KBs | Hazard magnitudes |
| | $D_i$ | KBs | KBs | Construction duration |
| | $\Delta Q^i(j)$ | – | IRP host, KBs | Wastage rate survey |
| | $Q_{sold}, Q_{bought}$ | – | Webfill host, KBs | Wastage rate survey |
| | undefined | – | EMS process③ and⑤ | Pollution and hazard survey |
| IRP host (Toolkit B) | $\Delta Q^i(j)$ | CPI host, KBs | – | Wastage rate survey |
| | $\Delta Q^i(j), C^i(j)$ | EMS process③, KBs | – | Reward |
| | $\Delta Q^i(j)$ | Webfill host, KBs | – | Quantity survey of waste |
| | $CPI_{waste}$ | – | CPI host, KBs | Wastage rate |
| | $Q^i_{de}(j), Q^i_{re}(j)$ | – | EMS process③ and④, KBs | Quantity survey of waste |
| | $Q_{sold}, Q_{bought}$ | – | Webfill host, KBs | Quantity survey of waste |

*Table 6.1* (Continued)

| Data host | Data name | Transfer to | Received from | Usefulness |
|---|---|---|---|---|
| Webfill host (Toolkit C) | $Q_{\text{sold}}, Q_{\text{bought}}$ | IRP host, KBs | – | Quantity survey of waste |
| | $Q_{\text{sold}}$ | EMS process④, KBs | – | Quantity survey of waste |
| | $Q_{\text{bought}}$ | EMS process③, KBs | – | Deliver to crews |
| | $Q_{\text{sold}}, Q_{\text{bought}}$ | CPI host, KBs | – | $CPI_{\text{waste}}$ |
| | Undefined | – | IRP host, KBs | Waste for exchange |
| | Undefined | – | EMS process④, KBs | Price of waste |
| EIA host | CPI, $CPI_i$ | – | CPI host, KBs | Data update for EIA report |
| | $\Delta Q^i(j)$ | – | IPR host, KBs | Data update for EIA report |
| | $Q_{\text{sold}}, Q_{\text{bought}}$ | – | Webfill host, KBs | Data update for EIA report |
| | Undefined | – | EMS process①, KBs | Data update for EIA report |
| | Undefined | – | EMS process②, KBs | Data update for EIA report |

Note
$CPI_{\text{waste}}$ represents the $CPI_i$ that involves waste impact only.

plays as a surrogate, a set of ontological commitments, a fragmentary theory of intelligent reasoning, a medium for efficient computation, and a medium of human expression (Davis *et al.* 1993). Leaving the conceptive discussions on the knowledge representation aside, the authors adopt two formats of CM knowledge to power the operation of the E+ system. They are the $CPI_i$ for the E+ EM Toolkit A and the $Q_i(j)_{\text{es}}$ for the E+ EM Toolkits B and C, which are the stochastic functions of several characters of construction processes (Chen and Li *et al.* 2004a,b). According to the definitions of the $CPI_i$ and the $Q_i(j)_{\text{es}}$, their values are computed by using statistic data and extracted by using ANN approach.

## 6.5    Experimental case studies

The experimental case studies conducted here combine data such as wastage at different construction stages (Vaid and Tanna 1997) from several separate cases (Chen and Li *et al.* 2000/4) and authors' experiences with a virtual construction project because there is no such construction project currently that has mature experience regarding the application of the E+ system as well as the inavailability of data at present to demonstrate the utilization of the E+ system from only one construction project. In this case, the aim of these experimental case studies focuses mainly on the utilization of the E+ model, and data adopted are for references only although there are practical backgrounds to support them. Therefore, it is necessary to note that as the prime objective of this case study is to demonstrate the usefulness of the E+ prototype, the authenticity of data adopted is de-emphasized. Case studies for real construction projects can be further conducted in the future when the E+ software environment has been realized.

### 6.5.1 Case study A

The experimental case study A presented in Table 6.2 demonstrates the process of the E+ model. The process of the EMS-based dynamic EIA in the experimental case study is divided into three stages corresponding to the construction cycle comprising pre-construction stage, construction stage, and post-construction stage (refer to Table 6.2). The EM-related data for the EMS-based dynamic EIA provided by the E+ system are different from stage to stage. At the pre-construction stage, there are two kinds of data for the EIA – the original set of $CPI_i$ and the $Q_{bought}$ requested by each crew; at the construction stage, there are four kinds of data for the EIA – the relay set of $CPI_i$, the $\Delta Q^i(j)$, the $Q_{sold}$, and the $Q_{bought}$; and at the post-construction stage, there are three kinds of data for the EIA – a final set of $CPI_i$, a total $Q_{sold}$, and a total $Q_{bought}$. The functions of current EM tools integrated in the E+ model are different, for example, the CPI approach deals with total adverse environmental impacts of construction processes, while the IRP approach and the Webfill approach deal with the C&D waste only, therefore the $CPI_i$ in this case study is thus represented by a $CPI_{waste}$ which represents the $CPI_i$ that involves waste impact only (refer to Table 6.1 and Equation 6.1).

Moreover, this experimental case study puts forward and utilizes the concepts of original $CPI_{waste}$, relay $CPI_{waste}$, and final $CPI_{waste}$ to demonstrate the process of the EMS-based dynamic EIA, and considers these three kinds of CPI an essential data in an EIA report. The original $CPI_{waste}$ means the $CPI_{waste}$ that is valued before a construction process, the relay $CPI_{waste}$ means the $CPI_{waste}$ that is devalued during a construction process, and the final $CPI_{waste}$ means the $CPI_{waste}$ that is finally valued after a construction process. Because the value of the $CPI_{waste}$ is regarded as an important data in an EMS-based EIA process, the changing process of the three kinds of $CPI_{waste}$ appropriately incarnates or reflects the process of a dynamic EIA. Thus an EMS-based dynamic EIA process is realized.

It is important to note that in order to value each $CPI_{waste}$, experts' experiences have to be used to set the magnitude of $h_i$ corresponding to changed amounts of $Q_{sold}$ and $Q_{bought}$. The expert experiences required to set $h_i$ are stored in the

Table 6.2 The E+ implementation for a dynamic EIA in a construction cycle: case study A

| Construction tasks | Duration (day) | EIA data at different construction stages | | | | | | | | |
| | | Pre-construction | | Construction | | | | Post-construction | | |
| | | Original CPI_waste | Q_bought (ton) | Relay CPI_waste | $\Delta Q^i(j)$ (ton) | Q_sold (ton) | Q_bought (ton) | Final CPI_waste | Total Q_sold (ton) | Total Q_bought (ton) |
|---|---|---|---|---|---|---|---|---|---|---|
| Caisson pile | 31 | 0.07 | 0.0 | 0.05 | 0.1 | 0.1 | 0.0 | 0.05 | 0.1 | 0.0 |
| Braced excavation | 57 | 0.13 | 0.0 | 0.15 | 0.5 | 0.3 | 0.0 | 0.10 | 0.5 | 0.0 |
| Transportation | 435 | 0.00 | 22.6 | 0.05 | 0.0 | 0.0 | 0.0 | 0.05 | 0.0 | 0.0 |
| Support system – building | 42 | 0.10 | 0.0 | 0.12 | 1.2 | 0.2 | 5.5 | 0.10 | 0.6 | 28.1 |
| Support system – demolition | 28 | 0.60 | 0.0 | 0.50 | 0.0 | 2.1 | 10.0 | 0.45 | 20.5 | 10.0 |
| Foundation construction | 43 | 0.10 | 0.1 | 0.20 | 3.3 | 3.2 | 1.1 | 0.18 | 5.6 | 1.2 |
| Structural RC work – rebar | 155 | 0.06 | 3.2 | 0.05 | 1.5 | 5.5 | 1.8 | 0.04 | 6.0 | 5.0 |
| Structural work – form | 155 | 0.36 | 5.7 | 0.30 | 1.2 | 0.2 | 1.1 | 0.28 | 0.5 | 6.8 |
| Structural work – concrete | 156 | 0.05 | 0.0 | 0.03 | 3.1 | 5.1 | 0.0 | 0.02 | 5.1 | 0.0 |
| Mansion work | 176 | 0.20 | 0.5 | 0.10 | 5.6 | 3.3 | 2.3 | 0.09 | 3.5 | 2.8 |
| Structural steel | 94 | 0.10 | 0.0 | 0.01 | 0.5 | 0.0 | 0.0 | 0.01 | 0.0 | 0.0 |
| Finish work – wall | 211 | 0.29 | 0.0 | 0.18 | 2.5 | 3.6 | 0.0 | 0.16 | 3.6 | 0.0 |
| Finish work – ceiling | 211 | 0.39 | 0.0 | 0.26 | 3.2 | 5.5 | 0.0 | 0.25 | 5.5 | 0.0 |
| Finish work – floor | 181 | 0.12 | 0.0 | 0.08 | 1.7 | 2.1 | 3.0 | 0.07 | 2.1 | 3.0 |
| Total | | 2.57 | 32.1 | 2.08 | | 31.2 | 24.8 | 1.85 | 53.6 | 56.9 |

Knowledge Re-use entity which is distilled from crude data in KBs including KB of environmental law and regulation, KB of environmental-friendly construction materials, KB of environmental-friendly construction machines, KB of environmental-friendly construction techniques, and KB of EM cases, etc. Therefore the EM-related construction knowledge effectively supports the process of EMS-based dynamic EIA.

The result of case study A indicates that the implementation of the E+ system can finally reduce the adverse environmental impacts of a construction project. For example, the total value of original $CPI_{waste}$ of all construction tasks in the experimental case study is 2.57, that of relay $CPI_{waste}$ is 2.08, and that of final $CPI_{waste}$ is 1.85. That is, the E+ system draws support from several EM tools such as the CPI approach, the IRP approach and the Webfill approach, and realizes an EMS-based dynamic EIA process, where the benefits of various EM tools can be shared within the E+ environment through EM-related data transfer and integrated data, information, and knowledge utilization.

### 6.5.2 Case study B

Being summarized in Table 6.3 and Figures 6.2 and 6.3, the experimental case study B demonstrates operation profiles of the E+ prototype by using quantified knowledge of each variable such as the $CPI_i$ and the $Q_i(j)_{es}$ etc. in correspondence with the re-use of CM knowledge for a dynamic EIA process in the construction lifecycle of a project. In addition to the $CPI_i$ and the $Q_i(j)_{es}$, a concept of CPI of C&D waste (denoted as $CPI_{i,waste}$ [CPI of waste of a specific construction operation $i$]) is introduced as a necessary complement of the CPI to each specific construction operation $i$ and a construction project. The authors further utilize the parameters of original $CPI_{i,waste}$, relay $CPI_{i,waste}$, and final $CPI_{i,waste}$ to demonstrate the process of the dynamic EIA in a construction lifecycle for evaluating the adverse environmental impacts of C&D waste, in parallel with a series of total CPI parameters including original $CPI_i$, relay $CPI_i$, and final $CPI_i$. The original $CPI_i/CPI_{i,waste}$ represents the $CPI_i/CPI_{i,waste}$ that is valued prior to a construction process for construction planning, the relay $CPI_i/CPI_{i,waste}$ denotes the $CPI_i/CPI_{i,waste}$ that is revalued during a construction process for construction pollution control, and the final $CPI_i/CPI_{i,waste}$ refers to the $CPI_i/CPI_{i,waste}$ that is finally valued after a construction process for knowledge re-use. Because the value of CPI is regarded as one important data in an EIA process, the changing process of the $CPI_i/CPI_{i,waste}$ can therefore reflect the dynamic alteration process of EIA/EM, which is enabled by the E+ system, and these two series of CPI values can be further used in an EM report and new construction projects.

It is necessary to note that in order to value the $CPI_i/CPI_{i,waste}$ of each construction process, experts' experiences have to be used according to the changed quantities of the $Q_i(j)_{es}$, the $Q_i(j)_{sd}$, and the $Q_i(j)_{bt}$, as well as the changed quantities of energy consumption. The expert experiences required to define the $CPI_i/CPI_{i,waste}$ are stored in the Knowledge Re-use entity, and are distilled from

Table 6.3 The E+ operation for a dynamic EIA process in a project construction lifecycle: case study B

| Construction process | Stage | Pre-construction | | | | Construction | | | | | Post-construction | | | | |
|---|---|---|---|---|---|---|---|---|---|---|---|---|---|---|---|
| | Duration (day) | Original CPI$_i$ (per day) | Original CPI$_{i,waste}$ (per day) | Q$_i$(j)$_{es}$ (ton/m²) | Q$_i$(j)$_{bought}$ (ton/m²) | Relay CPI$_i$ (per day) | Relay CPI$_{i,waste}$ (per day) | ΔQ$_i$(j) (ton/m²) | Q$_i$(j)$_{sold}$ (ton/m²) | Q$_i$(j)$_{bought}$ (ton/m²) | Final CPI$_i$ (per day) | Final CPI$_{i,waste}$ (per day) | Revised Q$_i$(j)$_{es}$ (ton/m²) | Final Q$_i$(j)$_{sold}$ (ton/m²) | Final Q$_i$(j)$_{bought}$ (ton/m²) |
| RC caisson pile | 31 | 0.65 | 0.07 | 7.60 | 0.00 | 0.60 | 0.05 | 0.02 | 0.01 | 0.00 | 0.60 | 0.05 | 7.57 | 0.01 | 0.00 |
| RC braced excavation | 57 | 0.20 | 0.13 | 3.80 | 0.00 | 0.35 | 0.15 | 0.03 | 0.01 | 0.00 | 0.30 | 0.10 | 3.76 | 0.01 | 0.00 |
| Transportation | 435 | 0.35 | 0.00 | N/A | N/A | 0.45 | 0.05 | N/A | N/A | N/A | 0.45 | 0.05 | 0.00 | N/A | 0.00 |
| Support system – building | 42 | 0.25 | 0.10 | 0.80 | 0.20 | 0.30 | 0.12 | 0.01 | 0.00 | 0.20 | 0.30 | 0.10 | 0.39 | 0.00 | 0.40 |
| Support system – demolition | 28 | 0.85 | 0.60 | N/A | 0.00 | 0.80 | 0.50 | 0.00 | 1.20 | 0.00 | 0.80 | 0.45 | N/A | 1.25 | 0.00 |
| RC foundation construction | 43 | 0.25 | 0.10 | 3.10 | 0.10 | 0.45 | 0.20 | 0.02 | 0.01 | 0.02 | 0.45 | 0.18 | 2.95 | 0.01 | 0.12 |
| Structural RC work – rebar | 155 | 0.15 | 0.06 | 0.60 | 0.60 | 0.15 | 0.05 | 0.01 | 0.01 | 0.02 | 0.15 | 0.04 | -0.04 | 0.01 | 0.62 |
| Structural RC work – form | 155 | 0.45 | 0.36 | 0.50 | 0.20 | 0.40 | 0.30 | 0.01 | 0.00 | 0.10 | 0.40 | 0.28 | 0.19 | 0.01 | 0.30 |
| Structural RC work – concrete | 156 | 0.65 | 0.05 | 0.60 | 0.00 | 0.60 | 0.03 | 0.01 | 0.01 | 0.00 | 0.60 | 0.02 | 0.58 | 0.01 | 0.00 |
| Concrete masonry work | 176 | 0.25 | 0.20 | 0.80 | 0.20 | 0.15 | 0.10 | 0.01 | 0.01 | 0.02 | 0.15 | 0.09 | 0.56 | 0.01 | 0.22 |
| Structural steel framework | 94 | 0.10 | 0.10 | 0.30 | 0.00 | 0.10 | 0.01 | 0.01 | 0.01 | 0.00 | 0.10 | 0.01 | 0.28 | 0.01 | 0.00 |
| Finish work – wall mortar | 211 | 0.35 | 0.29 | 0.10 | 0.00 | 0.25 | 0.18 | 0.01 | 0.01 | 0.00 | 0.25 | 0.16 | 0.08 | 0.01 | 0.00 |
| Finish work – ceiling mortar | 211 | 0.45 | 0.39 | 0.10 | 0.00 | 0.35 | 0.26 | 0.01 | 0.01 | 0.00 | 0.35 | 0.25 | 0.08 | 0.01 | 0.00 |
| Finish work – floor mortar | 181 | 0.25 | 0.12 | 0.10 | 0.00 | 0.15 | 0.08 | 0.01 | 0.01 | 0.01 | 0.15 | 0.07 | 0.07 | 0.01 | 0.01 |
| Total | | 5.20 | 2.57 | 18.40 | 1.30 | 5.10 | 2.08 | 0.16 | 1.30 | 0.37 | 5.05 | 1.85 | 15.27 | 1.36 | 1.67 |

Notes

CPI charts – refer to Figures 6.2 and 6.3.

1 The values of Q$_i$(j)$_{es}$, Q$_i$(j)$_{bought}$, and Q$_i$(j)$_{sold}$ are converted by using the quantities of various materials in each construction process.

2 The unit of area (m²) is based on the plot area of the building.

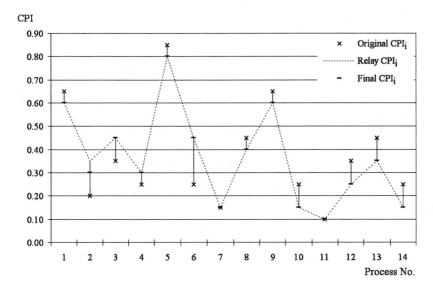

*Figure 6.2* CPI chart: case study B.

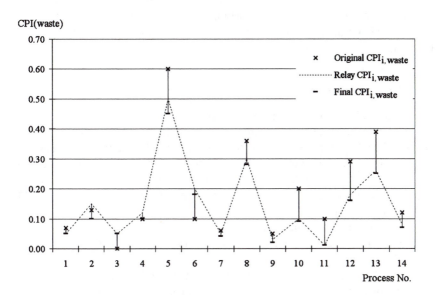

*Figure 6.3* CPI$_{waste}$ chart: case study B.

raw information in knowledge warehouse including knowledgebase of environmental law and regulation, knowledgebase of materials, equipment, techniques,
and EM cases, etc. All these procedures can finally effectively power the process
of knowledge-driven EMS-based dynamic EIA by reusing EM-related construction knowledge.

The result of this experimental case study indicates that the implementation
of the E+ system can finally reduce the adverse environmental impacts due
to construction. For example, the total value of original $CPI_i/CPI_{i,\text{waste}}$ of all
construction tasks is 5.20/2.57, the total value of relay $CPI_i/CPI_{i,\text{waste}}$ of all construction tasks is 5.10/2.08, and the total value of final $CPI_i/CPI_{i,\text{waste}}$ of all
construction tasks is 5.05/1.85; and there is a 3% reduction of CPI while there
is a 28% reduction of $CPI_{\text{waste}}$. Further to the reductions to the $CPI_i/CPI_{i,\text{waste}}$,
the authors provide two CPI charts (Figures 6.2 and 6.3), which can be used to
explain and analyse the changing process and the alterations of the $CPI_i/CPI_{i,\text{waste}}$
in each construction process. By using these results, the authors believe that
the operation of the E+ system can not only provide an integrated computer
tool to facilitate the implementation of a knowledge-driven EMS in construction projects but also create a decision-making environment to support further
analysis relating to the reduction of adverse environmental impacts due to
construction.

According to the results from case study B, it finally appears to the authors
that the E+ system can draw support from several mature EM tools such as
the CPI approach, the IRP approach, the Webfill approach, etc. and further
realize a knowledge-driven EMS-based dynamic EIA process, while the benefits
of various EM tools can be shared within an E+ system and EM-related data
transference and integrated data, information, and knowledge utilization can be
realized.

## 6.6 Future trends

Recommendations on the integrative knowledge management system for
environmental-conscious construction come from the usefulness, efficiency, and
benefit of the E+ prototype and EM tools, which have been demonstrated in this
chapter. However, due to the limitations of current research, it is recommended
to conduct further research on both the development of the E+ software environment and the development of more effective, efficient, and economical EM
tools for the E+ system.

First of all, the E+ model can be further developed to a Web-based environmental information and knowledge management system for contractors to implement EM in construction project management. According to the essential theory
and practice of EM in construction, environmental information is required in
construction planning, construction material management, C&D waste exchange,
etc., whilst EM knowledge in construction is essential to support decision-making
by using various EM tools. Because both environmental information and EM

knowledge are needed in the E+ system and the Internet is particularly suitable to implement effective and flexible CM by mobile site management units, the key issues in the development of such an E+ system are how to establish a Web-based software system and enable managers in different construction sites to use and share environmental information and knowledge on the same platform of E+, and how to capture, transfer, and re-use necessary data between the E+ system and current CM system. Moreover, additional functional components such as E+ EIA besides the E+ Toolkits are also under consideration.

Besides the development of the E+ system, further researches are also required in the development of fully user-oriented EM tools and their integration in the E+ system. The fully user-oriented EM tools can enable contractors or construction managers to use the EM toolkits easily by themselves without the help of tool developers. For example, the fully user-oriented tool of CPI can enable them to define the CPI of each construction process in a construction plan and then level the extremely high CPI, whilst the fully user-oriented tool of env.Plan (chen 2003) can enable them to transfer necessary data from construction plan alternatives to an ANP environment and thus to select the most environmental-friendly construction plan. In addition to the development of the fully user-oriented EM tools, improvements on the functionally different approaches for EM in construction focusing on the innovation of these approaches are also necessary. For example, the CPI of a construction process is defined by experts' experience currently, and this treatment is definitely practicable; however, in order to receive a wide recognition and minimize the arbitrary decision or subjective error on the definition of the CPI of each construction process, it is suggested to develop an objective calculation method to define the CPI of each possible construction process a contractor may use in construction planning. So both the development of the fully user-oriented EM tools and the consummation of current functionally different approaches for EM in construction are required in further researches.

Beyond the consummation of the E+ system and its components, additional functionally different approaches for EM in construction are also necessarily to be developed in order to improve the performance of the E+ system. Currently, the potential functionally different approaches to implementation of EM in construction include life-cycle analytical (LCA) approach and risk analytical approach for E+ EM Toolkit A, EIA template for new E+ EIA Toolkit, etc.

Although this research project has been accomplished with satisfied results, there are some limitations not only within the research but also in the duplicate implementation of the EM tools developed in the research. The limitations of the research exist in the following two areas:

- This research has not accomplished an E+ software environment to further demonstrate its usefulness, and efficiency in EM in construction.
- The CPI method has not been developed into a fully user-oriented tool that can help contractors to deal with any CPI problem in construction planning.

As a conclusion, it is recommended that further research and development for the E+ system focus on the development of a Web-based E+ system, the consummation and innovation of current EM tools in construction, and the development of new EM approaches for the E+ system.

## 6.7   Conclusions

This chapter presents a research for an integrative methodology named E+ for EMS-based dynamic EIA in construction, which integrates various EM approaches with a general EMS process throughout all construction stages in a construction project. The EM approaches integrated in the E+ are divided into three categories: EM Toolkit A for pre-construction, EM Toolkit B for construction, and EM Toolkit C for post-construction. These EM Toolkits are further integrated with ISO 14001 EMS and EM Knowledge Capture and Re-use entities for an integrative knowledge management system for environmental-conscious construction. In addition to the proposed E+ prototype, an experimental case study has also been conducted to demonstrate the usefulness and efficiency of the E+ system. The E+ is expected to effectively, efficiently, and economically assist contractors to enhance their EM techniques and environmental performances in construction project management, and to overcome the weakness of static EIA, formally applied in the construction industry in some countries, by the dynamic EIA process, where the necessary data for an EIA report can be updated in the construction cycle.

Regarding the integrative methodology of knowledge management system for EM in construction, this chapter mainly contributes to existing theory for EM in construction in the area of quantitative analytical approaches and their integrative implementation. According to the literature review and questionnaire survey for this research, the lack of an effective, efficient, and economical quantitative analytical approach is one of the obstacles to implementing EM in construction. Therefore, there are four points of contributions from this research to the existing theory or practice for EM in construction:

- This research has developed an integrative methodology (E+) for implementing EMS and knowledge management in construction, with a rigorous dynamic EIA model based on various functionally different approaches to EM in a construction cycle. The E+ prototype is originally created in both theory and practice for EM in construction, and it is open to further integration of various functionally different approaches for EM in construction other than the three EM tools presented in this chapter. Because the E+ is both EMS-oriented and process-oriented in construction, it can help contractors to implement EM from a messy situation to a normal system and to effectively share EM knowledge and information internally and externally.
- The CPI method integrated in the E+ model is a quantitative approach to predicting and levelling complex adverse environmental impacts potentially

generated from construction and transportation due to the implementation of a construction plan. As a result, the CPI method has been integrated into E+ EM Toolkit A, one functional section of E+ system, to carry out the task in environmental-conscious construction planning.

- The IRP method is a quantitative approach to reduce wastage of construction materials on a construction site, and it is designed to be effectively implemented by using a bar-code system. The IRP is then integrated in the E+ EM Toolkit B, another functional section of E+, as a basic component.
- The Webfill method is an e-commerce model designed for the trip-ticket system to effectively reduce, re-use, and recycle C&D waste. Although there is lack of data to prove the efficiency in reality, the computer simulation results, and a questionnaire survey from another research (Chen 2003) have proved that the Webfill system can effectively realize the design function. As a result, the Webfill is also integrated in the E+ model as an important component of the E+ EM Toolkit C.

Although the software environment of the E+ has not been presented in this chapter, the demonstration of the E+ system in the experimental case study enabled a closer understanding of how the E+ system can be effectively applied for EM in construction, and it also unveiled that the E+ methodology is flexible in the integrative implementation of functionally different quantitative approaches to EM in construction. In order to promote the implementation of the E+ model, further research is required to transfer the E+ model to a computer software environment and improve current EM tools and develop more EM approaches as subsidiary components of the E+ system to deal with all adverse environmental impacts of construction for total EM in construction project management.

# A questionnaire about EMS application

## A.1  A covering letter

March 31, 2001

Subject: A questionnaire on the adoption and implementation of the ISO 14000 series of standards and environmental management systems in construction enterprises in mainland China.

Dear Sir or Madam,

I am a PhD candidate in the department of Building and Real Estate in the Hong Kong Polytechnic University, and I am studying on environmental management in construction projects in China. I submit this questionnaire to you personally, and I will appreciate your attention, cooperation, response, and comments.

Ever since the ISO 14000 series was introduced in 1996, more and more attention has been paid to environmental management system in construction industry globally, and has become the hot spot in construction management since the ISO 9000 series introduced in 1992. In Hong Kong, there are already 21 construction companies who have obtained ISO 14001 EMS certificates by the end of March 2001. We believe that more and more construction companies in the mainland of China will adopt and implement EMS, including the ISO 14000 series EMS, for their sustainable development in a society where environment is a concern. The aim of this questionnaire survey is to find out the degree of self-identification with the ISO 14000 series and EMS in some large-scale construction companies in selected cities of mainland China. The results of this survey can provide valuable data to my research and dissertation – quantitative analytical approach for environmental management in construction projects.

Your comprehension and sustentation are great affirmation and assistance to my research! I frankly assure that all information about your company you provide in this questionnaire survey will only be used in statistical analysis in this learning research, and I will never open any individual information to the public. Kindly take time to complete this questionnaire, and try to send it back to me as soon as possible.

Please leave your contact information at the end of this questionnaire if you want to see the report of this survey. I will send the entire survey report to you later. In case you have no time to do this survey, could you please transfer this questionnaire to your reliable colleagues? If it is possible, could you please pass on this questionnaire to more colleagues of yours? I am now in Chengdu, and will go back to Hong Kong in August. Please call me at 028-5572374 during these days for anything I should do.

Thank you very much for your comprehension, assistance, and support!

Sincerely yours,

(Signature)
Zhen Chen
PhD Candidate
Address:
TU410, Department of Building and Real Estate
The Hong Kong Polytechnic University, Hong Kong
Tel: 852-27665873, Fax: 852-27645131
E-mail: z.chen@polyu.edu.hk
URL: http://hk.geocities.com/at55379/index.html

## A.2  Questionnaire

**Part 1   Background** (Please check all that applies)

1.1   Major source of construction projects for your company:

☐   Governmental Project _____%
☐   Public Project _____%
☐   Private Project _____%

1.2   Major types of construction projects undertaken by your company and their normal percentage:

☐   National Civil Project _____%
☐   Local Civil Project _____%
☐   Industrial Project _____%
☐   Commercial Project _____%
☐   Residential Project _____%
☐   Electrical Works _____%
☐   Water Supply–Drainage Works _____%

- ☐ Heating and Ventilating Works _____%
- ☐ Gas Supply Works _____%
- ☐ Others _____%

1.3 Total annual contracts of your company in 2000 is _____ million dollars (RMB), and normal annual contracts is _____ million dollars (RMB).

1.4 Total expenditure for environmental management (EM) of your company in 2000 is _____ dollars, and normal annual expenditure is _____ dollars.

1.5 There are _____ administrators in your company, and _____ of them are involved in EM.

1.6 There are _____ subcontractors working with your company now, and normally there are _____ subcontractors.

1.7 There are _____ subcontractors working with your company who have ISO 14001 accreditations.

1.8 There are _____ material and machine suppliers for your company, and normally there are _____ suppliers.

1.9 There are _____ suppliers for your company who have ISO 14001 accreditations.

1.10 State of ISO 14001 certification for your company:

- ☐ Registered. Requested at year _____, obtained at year _____, and it took about _____ months.
- ☐ Under assessment. Requested at year _____, will obtain at year _____, and it took about _____ months.
- ☐ Failed. Requested at year _____, failed at year _____, and it took _____ months.
- ☐ Registered but expired. Registered at year _____, took _____ months, and expired at year _____.
- ☐ Have not applied for, but preparing to. Will request at year _____.
- ☐ Do not want to apply. (Please ignore questions 1.11–1.14.)

1.11 Total expense for ISO 14001 certification of your company is _____ million dollars.

1.12 Total acceptable expense for adoption and implementation of environmental management system based on ISO 14001 in your company: (Unit is Chinese dollars (RMB))

- ☐ Free
- ☐ Less than 0.1 million
- ☐ 0.10–0.25 million
- ☐ 0.25–0.50 million
- ☐ 0.5–0.75 million
- ☐ more than 0.75 million

1.13   Reasons for applying for ISO 14001 registration of your company:

| Potential reasons | Grade of influence (influence increase by degrees from 1 to 10) | | | | | | | | | |
|---|---|---|---|---|---|---|---|---|---|---|
| | 1 | 2 | 3 | 4 | 5 | 6 | 7 | 8 | 9 | 10 |
| International competitive pressure | | | | | | | | | | |
| National competitive pressure | | | | | | | | | | |
| Management standardization | | | | | | | | | | |
| Improving document-handling in company | | | | | | | | | | |
| Reduce project cost | | | | | | | | | | |
| Obey laws and regulations | | | | | | | | | | |
| Client's request | | | | | | | | | | |
| Conscious of environmental protection | | | | | | | | | | |
| Others (Please specify) | | | | | | | | | | |

1.14   Main considerations while applying for ISO 14001 registration for your company:

| Considerations | Grade of influence (influence increase by degrees from 1 to 10) | | | | | | | | | |
|---|---|---|---|---|---|---|---|---|---|---|
| | 1 | 2 | 3 | 4 | 5 | 6 | 7 | 8 | 9 | 10 |
| Cost of implementation | | | | | | | | | | |
| Investment of additional human resource | | | | | | | | | | |
| Burden of EM document | | | | | | | | | | |
| Combination with other management works | | | | | | | | | | |
| Restructure organization | | | | | | | | | | |
| Profits | | | | | | | | | | |
| Others (Please specify) | | | | | | | | | | |

1.15   Potential benefits of ISO 14001 registration for your company:

| Potential benefits | Grade of benefits (benefits increase by degrees from 1 to 10) | | | | | | | | | |
|---|---|---|---|---|---|---|---|---|---|---|
| | 1 | 2 | 3 | 4 | 5 | 6 | 7 | 8 | 9 | 10 |
| Enlarge occupation in market | | | | | | | | | | |
| Increase productivity | | | | | | | | | | |
| Decrease harm to environment | | | | | | | | | | |
| Increase company's profits | | | | | | | | | | |
| Improve trading and public status | | | | | | | | | | |
| Fulfill clients' demands | | | | | | | | | | |
| Others (Please specify) | | | | | | | | | | |

1.16   Extent of desire of applying for ISO 14001 certification of your company:

- ☐   Strongly reject
- ☐   Reject, but can be considered
- ☐   Not decided yet
- ☐   Accept, but need consideration
- ☐   Strongly accept

1.17   Will your company apply for an update of ISO 14001 registration in the future?

- ☐   Definitely not
- ☐   May not
- ☐   Unsure
- ☐   Maybe
- ☐   Definitely yes

1.18   Why does your company not think of applying for an ISO 14001 certificate?

- ☐   Higher cost
- ☐   No interest
- ☐   Small profits
- ☐   Not necessary
- ☐   Lack of professionals

☐  Conflicts in organization
☐  Lesson from failure on EM
☐  Others (Please specify)

**Part 2  Adoption and implementation of ISO 14001 EMS and internal EMS**
(Please check all that applies)

2.1  Do you have any other internal EMS except ISO 14001 EMS in your company?

☐  Have
☐  Not have (Please skip to question 2.5)
☐  Under construction
☐  Failed

2.2  Your internal EMS is

☐  Internal total quality and environmental management system
☐  Internal EMS
☐  Others (Please specify)

2.3  Do you implement EM according to ISO 14001 EMS and internal EMS?

☐  Both ISO 14001 EMS and ISO 9000 QMS
☐  ISO 14001 EMS only
☐  Internal TQEMS only
☐  Internal EMS only
☐  ISO 14001 EMS and Internal TQEMS
☐  ISO 14001 EMS and Internal EMS
☐  Others

2.4  The EM in your company focuses on

☐  Energy efficiency
☐  Control and reduction of quantity of waste
☐  Control and reduction of noise
☐  Control and reduction of air pollution
☐  Recycling materials and equipment
☐  Recycling waste materials and packing
☐  Control and reduction of accident
☐  Control and reduction of geological deformation
☐  Others (Please specify)

2.5   Your opinion about the pertinent factors on environmental impact of construction projects

| Pertinent factors | Grade of pertinence (pertinence increase by degrees from 1 to 10) | | | | | | | | | |
|---|---|---|---|---|---|---|---|---|---|---|
| | 1 | 2 | 3 | 4 | 5 | 6 | 7 | 8 | 9 | 10 |
| Governmental (national and local) laws and statutes | | | | | | | | | | |
| Governmental civilians' attitude on executing the laws | | | | | | | | | | |
| ISO 14001 registration | | | | | | | | | | |
| Subcontractors' attitude towards cooperation | | | | | | | | | | |
| Interior collectivism | | | | | | | | | | |
| Exterior competitive pressure | | | | | | | | | | |
| Requirements from clients | | | | | | | | | | |
| Company's profit | | | | | | | | | | |
| Company's consciousness of environmental protection | | | | | | | | | | |
| Construction technologies | | | | | | | | | | |
| Construction materials and machines | | | | | | | | | | |
| Others (Please specify) | | | | | | | | | | |

## Part 3   Advantages and disadvantages of adoption and implementation of the ISO 14000 series

3.1   Advantages of adoption and implementation of the ISO 14000 series are that it can

| Advantage | Grade of advantage (advantage increases by degrees from 1 to 10) | | | | | | | | | |
|---|---|---|---|---|---|---|---|---|---|---|
| | 1 | 2 | 3 | 4 | 5 | 6 | 7 | 8 | 9 | 10 |
| Increase occupancy on the market | | | | | | | | | | |
| Enhance visualization and celebrity | | | | | | | | | | |

| Advantage | Grade of advantage (advantage increases by degrees from 1 to 10) | | | | | | | | | |
| --- | --- | --- | --- | --- | --- | --- | --- | --- | --- | --- |
| | 1 | 2 | 3 | 4 | 5 | 6 | 7 | 8 | 9 | 10 |
| Increase profit | | | | | | | | | | |
| Improve productivity | | | | | | | | | | |
| Improve quality | | | | | | | | | | |
| Decrease duration of a construction project | | | | | | | | | | |
| Protect environment | | | | | | | | | | |
| Increase the degree of clients' satisfaction | | | | | | | | | | |
| Improve the efficiency of EM | | | | | | | | | | |
| Others (Please specify) | | | | | | | | | | |

3.2   Disadvantages of adoption and implementation of the ISO 14000 series are

| Disadvantage | Grade of disadvantage (disadvantage increases by degrees from 1 to 10) | | | | | | | | | |
| --- | --- | --- | --- | --- | --- | --- | --- | --- | --- | --- |
| | 1 | 2 | 3 | 4 | 5 | 6 | 7 | 8 | 9 | 10 |
| Cost of registration | | | | | | | | | | |
| Preparative work in earlier stage | | | | | | | | | | |
| Cost of implementation | | | | | | | | | | |
| Change of processes of project management | | | | | | | | | | |
| Additional human resource | | | | | | | | | | |
| Change of internal EMS | | | | | | | | | | |
| Change of internal organizational structure | | | | | | | | | | |
| Others (Please specify) | | | | | | | | | | |

## Part 4    Perceptions of the ISO 14000 series of standards and EMS

4.1    Perceptions of the ISO 14000 series of standards on different administrative levels

| Internal administrative level | Grade of perception (perception increases by degrees from 1 to 10) | | | | | | | | | |
|---|---|---|---|---|---|---|---|---|---|---|
| | 1 | 2 | 3 | 4 | 5 | 6 | 7 | 8 | 9 | 10 |
| Worker | | | | | | | | | | |
| Common employee | | | | | | | | | | |
| Junior manager | | | | | | | | | | |
| Project manager | | | | | | | | | | |
| Senior manager | | | | | | | | | | |

4.2    Perceptions of EMS on different administrative levels

| Internal administrative level | Grade of perception (perception increases by degrees from 1 to 10) | | | | | | | | | |
|---|---|---|---|---|---|---|---|---|---|---|
| | 1 | 2 | 3 | 4 | 5 | 6 | 7 | 8 | 9 | 10 |
| Worker | | | | | | | | | | |
| Common employee | | | | | | | | | | |
| Junior manager | | | | | | | | | | |
| Project manager | | | | | | | | | | |
| Senior manager | | | | | | | | | | |

## Part 5    Some correlated questions about the ISO 14000 series and EM

5.1    Your opinion about the statement, "The ISO 14000 series is necessary and important for your company to adopt and implement EMS"

☐    Strongly agree
☐    Agree
☐    Neutral
☐    Disagree
☐    Strongly disagree

5.2   Your opinion about the statement, "The ISO 14000 series is contributive to your company in improving EM"

☐   Strongly agree
☐   Agree
☐   Neutral
☐   Disagree
☐   Strongly disagree

5.3   Your opinion about the statement, "EMS is essential for a construction company to improve EM"

☐   Strongly agree
☐   Agree
☐   Neutral
☐   Disagree
☐   Strongly disagree

5.4   Your opinion about the statement, "It is necessary to implement internal EMS and adopt ISO 14000 at the same time"

☐   Strongly agree
☐   Agree
☐   Neutral
☐   Disagree
☐   Strongly disagree

5.5   Your opinion about the statement, "The cost of EM is important than EM itself"

☐   Strongly agree
☐   Agree
☐   Neutral
☐   Disagree
☐   Strongly disagree

5.6   Your opinion about the statement, "Activity-costing control can be a good tool for managing the cost of EM"

☐   Strongly agree
☐   Agree
☐   Neutral
☐   Disagree
☐   Strongly disagree

5.7   Your opinion about the statement, "Similar to the use of air pollution index to evaluate air quality in cities, construction pollution index (CPI) of an activity can be used to evaluate environmental impact of a construction activity. The CPI can be an efficient approach for EM through the control of activities' CPI and, by implication, the project's CPI under an acceptable cost"

- ☐ Strongly agree
- ☐ Agree
- ☐ Neutral
- ☐ Disagree
- ☐ Strongly disagree

5.8   Your opinion about the statement, "It is important for a contractor to consider about the environmental impact of materials when he wants to select a supplier, similarly, it is important for a contractor to consider about the implementation of EMS in construction when he wants to select a subcontractor"

- ☐ Strongly agree
- ☐ Agree
- ☐ Neutral
- ☐ Disagree
- ☐ Strongly disagree

5.9   Your opinion about the statement, "Waste and second-hand materials and equipment can be traded by using an exchange platform/portal on the Internet, and then the total waste from the construction industry can be reduced. So the electronic commerce firm can be a commercial associate with construction companies on their EMS"

- ☐ Strongly agree
- ☐ Agree
- ☐ Neutral
- ☐ Disagree
- ☐ Strongly disagree

## Part 6 Potential influences on adopting the ISO 14000 series in construction companies

| Potential influential reasons | Grade of influence (influence increases by degrees from 1 to 10) | | | | | | | | | |
|---|---|---|---|---|---|---|---|---|---|---|
| | 1 | 2 | 3 | 4 | 5 | 6 | 7 | 8 | 9 | 10 |
| Governmental requirement on adopting the ISO 14000 series in construction industry | | | | | | | | | | |
| Governmental encouragement on financial subsidies, e.g. tax deduction/return | | | | | | | | | | |
| Governmental encouragement on non-financial allowance | | | | | | | | | | |
| Competitive pressure from international construction industry within WTO | | | | | | | | | | |
| Competitive pressure from domestic construction industry | | | | | | | | | | |
| Cost of implementation of the ISO 14000 series EMS (About RMB 0.3M) | | | | | | | | | | |
| Cost of ISO 14001 registration (About RMB 50,000) | | | | | | | | | | |
| Internal initiative consciousness on implementation of EMS | | | | | | | | | | |
| Requirement and pressure from clients or suppliers | | | | | | | | | | |
| Expectation from clients or suppliers | | | | | | | | | | |
| Introducing the ISO 14000 series for establishing enterprise's internal EMS | | | | | | | | | | |
| Encouraging subcontractors to adopt ISO 14001 for improving the level of EM | | | | | | | | | | |

(Continued)

| Potential influential reasons | Grade of influence (influence increases by degrees from 1 to 10) | | | | | | | | | |
|---|---|---|---|---|---|---|---|---|---|---|
| | 1 | 2 | 3 | 4 | 5 | 6 | 7 | 8 | 9 | 10 |
| Various additional EM documents on adopting ISO 14001 | | | | | | | | | | |
| Interruption and adjustment of construction processes on implementing ISO 14001 | | | | | | | | | | |
| Entire employees' training and education before implementing ISO 14001 EMS | | | | | | | | | | |
| Additional cost on training functionaries inside company | | | | | | | | | | |
| The necessity of management involvement on adopting ISO 14001 | | | | | | | | | | |
| Cost of ISO 14001 EMS assessment, certification, and maintenance | | | | | | | | | | |
| Additional cost of human resource on adopting and implementing ISO 14001 | | | | | | | | | | |
| Additional cost of reorganization on adopting and implementing ISO 14001 | | | | | | | | | | |
| Lack of reliable consulting companies on tutorship of adoption of ISO 14001 | | | | | | | | | | |
| Additional cost of failure on adopting ISO 14001 EMS | | | | | | | | | | |

## Part 7 Potential influences on implementing the ISO 14000 series in construction companies

| Potential influential reasons | Grade of influence (influence increases by degrees from 1 to 10) | | | | | | | | | |
|---|---|---|---|---|---|---|---|---|---|---|
| | 1 | 2 | 3 | 4 | 5 | 6 | 7 | 8 | 9 | 10 |
| Additional cost of implementation of ISO 14001 EMS | | | | | | | | | | |
| Employees' attitude towards cooperation on implementation of ISO 14001 | | | | | | | | | | |
| Administrators' attitude towards cooperation on implementation of ISO 14001 | | | | | | | | | | |
| Subcontractors' attitude towards cooperation on implementation of ISO 14001 | | | | | | | | | | |
| Suppliers' attitude towards cooperation on implementation of ISO 14001 | | | | | | | | | | |
| Impacts and additional expense of interruption and adjustment on construction | | | | | | | | | | |
| Success or failure in adjustment of organizational structure inside the enterprise | | | | | | | | | | |
| Success or failure in combination with other EMSs inside the enterprise | | | | | | | | | | |
| Success or failure in the maintenance and continuous assessment of ISO 14001 EMS | | | | | | | | | | |
| Success or failure in employees' training and education inside the enterprise | | | | | | | | | | |
| Success or failure in administrator's training and education inside the enterprise | | | | | | | | | | |

(Continued)

| Potential influential reasons | Grade of influence (influence increases by degrees from 1 to 10) | | | | | | | | | |
|---|---|---|---|---|---|---|---|---|---|---|
| | 1 | 2 | 3 | 4 | 5 | 6 | 7 | 8 | 9 | 10 |
| Lack of pressure from the government | | | | | | | | | | |
| Lack of pressure from the clients | | | | | | | | | | |
| Lack of pressure from the competitors inside construction industry | | | | | | | | | | |
| No competitors implement the ISO 14000 series first inside the construction industry | | | | | | | | | | |
| High expense on implementation | | | | | | | | | | |
| Multifarious documental operation process of ISO 14001 | | | | | | | | | | |
| Lack of suitable technology and material for environmental protection | | | | | | | | | | |
| Applicability of the ISO 14000 series in construction enterprises | | | | | | | | | | |
| Correspondence and cooperation of design and construction | | | | | | | | | | |

**Part 8 Additional comments**

Thank you for your participation! For further contacts, please provide the following information:

---

Company Name:

---

Website:

---

Business Scope:

---

Grade:

---

Contact Person:

---

Position:

---

Mail Address:

---

City/Province:                                        Postcode:

---

Telephone:                                             E-mail:

---

# A decision-making model

## B.1 Introduction

The analysis presented in Chapter 2 identified that there are five classes (critical factors) directly affecting the acceptability of the ISO 14000 series in the Shanghai construction industry. In this appendix, the five critical factors are integrated into a decision-making model which can assess whether a contracting company is positive or negative to the acceptance of the ISO 14000 series. In addition, the model can also enable contractors to identify weak aspects in adopting and implementing the ISO 14000 series, assuming that they are willing to accept the ISO 14000 series.

Discriminant analysis is used to develop the decision-making model as it is useful in situations where one wants to build an evaluation model of group membership based on observed characteristics of each case and an established predictive model can then be applied to new cases with measurements for the predictor variables for unknown group membership (Norusis 2000). Such an analysis method has also been used to develop an evaluation model for a company's decision to adopt ISO 14001 in Singapore (Quazi *et al.* 2001).

There are two basic requirements in using the discriminant analysis in statistical inferences: one is that the independent variables obey a normal distribution, another is that the independent variables are linearly related to the dependent variable (Norusis 2000). The procedure of a discriminant analysis generates a discriminant function based on linear combinations of the normally distributed predictor variables that provide the best discrimination between the contractors who are either positive or negative to acceptance of the ISO 14000 series. Therefore, before using the discriminant analysis, it is necessary to ensure that the five critical factors satisfy the two basic requirements.

Let us assume that the five critical factors are independent variables and are represented by $C_{5C}$ (Classes of 5C), where $C_{comd}$, $C_{cond}$, $C_{comp}$, $C_{coop}$, and $C_{cost}$ represents the $\underline{C}$lasses of the governmental $\underline{C}$ommand-and-control regulations, the technology $\underline{C}$onditions, the $\underline{C}$ompetitive pressures, the $\underline{C}$ooperative attitude, and the $\underline{C}$ost–benefit efficiency, respectively. The acceptability of the ISO 14000 series is a dependent variable and is represented by $A_{ISO\ 14k}$ (e.g. for accepters, $A_{ISO\ 14k} = 1$; for others, $A_{ISO\ 14k} = 0$).

## B.2 Probability Distributions of the $C_{5C}$

A normal probability plot, e.g. Q-Q (quantile-quantile) plot, is generally used to check whether variables follow a normal distribution when one wants to assess normality (Norusis 2000). In the quantitative analysis of the survey data, the Q-Q plot is applied to identify and assess normality of each class of the $C_{5C}$. The finished Q-Q plots, as shown in Figure B.1, indicate that all five classes, with observed significance levels approximately below 0.01 in the Kolmogorov-Smirnov statistical tests of normality, are normally distributed. This is because if the sample is from a normal distribution, points will cluster around a straight line in a Q-Q plot and if the observed significance level is small enough, usually less than 0.05 or 0.01, the null hypothesis is rejected (Norusis 2000).

## B.3 Linear relationships between the $C_{5C}$ and the $A_{ISO\ 14k}$

The linear relationships between the $C_{5C}$ and the $A_{ISO\ 14k}$ can be measured by both quantitative indices and graphic matrix in the SPSS®. The quantitative indices used are normally tolerance and variance inflation factor (VIF). The tolerance is a statistical value used to determine how much the independent variables are

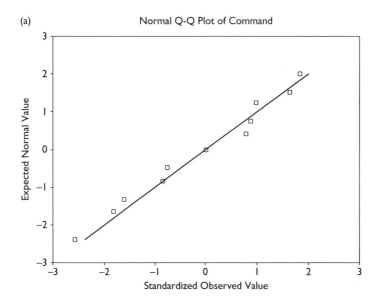

*Figure B.1* Normal Q-Q plots of the $C_{5C}$ with total 72 respondents. Normal Q-Q plots of Class 1: governmental regulations; Normal Q-Q plots of Class 2: technology conditions; Normal Q-Q plots of Class 3: competitive pressures; Normal Q-Q plots of Class 4: cooperative attitude; Normal Q-Q plots of Class 5: Cost–benefit efficiency.

(b)

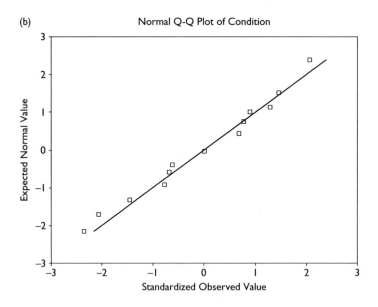

(c)

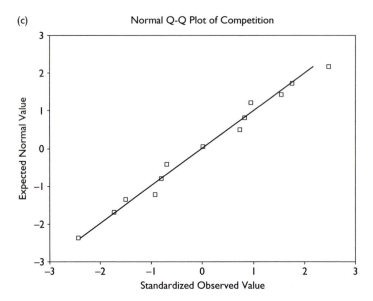

*Figure B.1* (Continued).

(d)

Normal Q-Q Plot of Cooperation

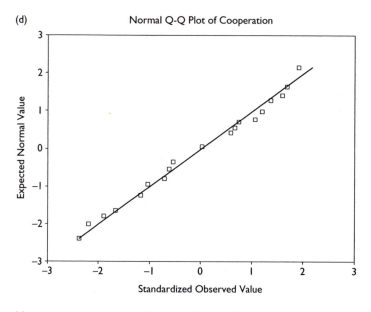

(e)

Normal Q-Q Plot of Cost

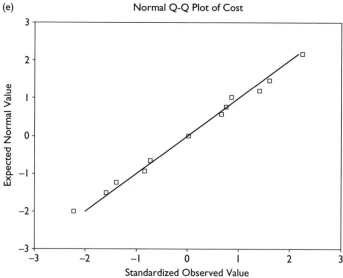

Notes

One-Sample Kolmogorov-Smirnov Test (Test distribution is Normal)

class 1: Governmental regulations: Significance = 0.000;

class 2: Technology conditions: Significance = 0.007;

class 3: Competitive pressures: Significance = 0.000;

class 4: Cooperative attitude: Significance = 0.000;

class 5: Cost–benefit efficiency: Significance = 0.000.

*Figure B.1* (Continued).

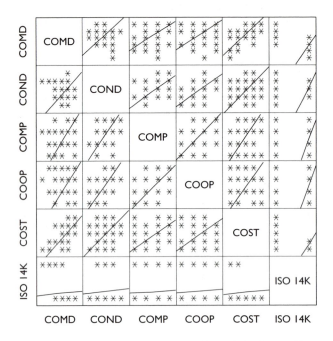

*Figure B.2* Collinearity statistics of the $C_{5C}$ and the $A_{ISO\ 14k}$ (Scatterplot matrix).

linearly related to one another (multicollinear) and a variable with very low toler-ance contributes little information to a model and thus brings noise to the resultant model; and the VIF is a reciprocal of the tolerance, and a large VIF value is an indicator of multicollinearity (Norusis 2000). The calculated results of these two indices are listed in Figure B.2. Since no value of VIF exceeds 10, it can be con-cluded that values of both tolerance and VIF indicate inconsequential coollinerity between the $C_{5C}$ (Field 2000; Quazi *et al.* 2001). On the other hand, the graphic matrix, such as a scatterplot matrix, can be used to check for linearity of variances by plotting any two of the dependent variables. The scatterplot matrix of the $C_{5C}$ and the $A_{ISO\ 14k}$ (refer to Figure B.2) suggests that although all of the inde-pendent variables have inconsequential linear relationships with the dependent variable $A_{ISO\ 14k}$, fit line can be drawn through some fit methods, such as a linear regression.

## B.4 Discriminant function for $C_{5C}$ and $A_{ISO\ 14k}$

Based on the discriminant analysis with $C_{5C}$ and $A_{ISO\ 14k}$ using the SPSS®, a mul-tiple linear regression equation that predicts the ISO 14000 series' acceptability

*Table B.1* Discriminant function coefficients for linear acceptability evaluation model

|  | Canonical discriminant function | Classification function coefficients (Fisher's linear discriminant functions) | |
| --- | --- | --- | --- |
|  | Standardized | $A_{ISO\ 14k} = 0$ | $A_{ISO\ 14k} = 1$ |
| $\alpha_0$ | 0.000 | −108.470 | −84.170 |
| $\alpha_{comd}$ | +1.026 | 9.383 | 5.850 |
| $\alpha_{cond}$ | −0.112 | 4.882 | 5.203 |
| $\alpha_{comp}$ | −0.274 | 5.368 | 6.224 |
| $\alpha_{coop}$ | −0.043 | 1.194 | 1.298 |
| $\alpha_{cost}$ | +0.302 | 5.430 | 4.574 |

*Table B.2* Classification results of the linear acceptability evaluation model

| Original actual group | Sample size | Predicted group of non-accepters | Predicted group of accepters |
| --- | --- | --- | --- |
| Group of non-accepters | 58 | 54 | 4 |
| Group of accepters | 14 | 4 | 10 |

Note
88.9% of originally grouped cases are correctly classified using Equation B.1.

is developed in Equation B.1, and all coefficients in Equation B.1 are presented in Table B.1.

$$A_{ISO\ 14k} = \alpha_0 + \alpha_{comd}\,\overline{C}_{comd} + \alpha_{cond}\,\overline{C}_{cond} + \alpha_{comp}\,\overline{C}_{comp} + \alpha_{coop}\,\overline{C}_{coop}$$
$$+ \alpha_{cost}\,\overline{C}_{cost} \tag{B.1}$$

where $\alpha_i$ represents coefficients, and $\overline{C}_i$ represents the average score of each class.

Using Equation B.1 to run through the whole sample population, the linear discriminant model (Equation B.1) correctly classified 88.9% of the companies into the two groups (refer to Table B.2). This percentage is within the range specified by Quazi *et al.* (2001), which indicates that the discriminant model is useful.

## B.5  Validation

The linear discriminant function (Equation B.1) has also been validated by using data from our follow-up questionnaire (see Appendix A) surveys conducted among main contractors in five representative cities in mainland China, where the average EIA rate was 85% and the average ISO 14001 registration rate was 0.06%. The validation results indicate that the total rate of correct classification with Equation B.1 is as low as 78% (refer to Table B.3), and the highest correct match rate occurs with samples from Shanghai, where Equation B.1 is used.

As a result, because the inconsequential co-linearity between both of the independent variables and the dependent variable surpasses the boundary of the assumption of a discriminant analysis in linear regression, the pure linear predictive model is almost rejected not only in this research but also in another (Quari *et al.* 2001) with a similar correct classification rate. Moreover, our further attempts in making a multiple nonlinear acceptability evaluation model can provide a correct classification rate higher than 88.9% for the time being, and it indicated that a multiple nonlinear regression equation is necessary.

## B.6  Model application

The linear discriminant model can be applied to assist contractors to make decisions on whether to adopt and implement the ISO 14000 series. The model can also be used to identify weak aspects of a contracting company in adopting and implementing the ISO 14000 series, assuming the company accepts the ISO 14000. The reasoning mechanism of the model can be expressed as follows:

**FOR** $i = 1$ to 5
**IF** $S_1 > \overline{C}_1$ or $S_2 < \overline{C}_2$ or $S_3 < \overline{C}_3$ or $S_4 < \overline{C}_4$ or $S_5 > \overline{C}_5$
**THEN** display $Sug_i$
**OTHERWISE** display *Congratulations! Your company is ready to have ISO 14001 accreditation.*

*Table B.3* Revaluation results of the linear acceptability evaluation model

| City name | Sample size | Correct match | Wrong match | Rate of correct classification (%) |
|---|---|---|---|---|
| Shanghai | 20 | 18 | 2 | 90 |
| Tsingtao | 20 | 16 | 4 | 80 |
| Jinan | 20 | 15 | 5 | 75 |
| Chengdu | 20 | 14 | 6 | 70 |
| Chongqing | 20 | 15 | 5 | 75 |
| Total | 100 | 78 | 22 | 78 |

*Table B.4* Checklist for decision-making on acceptance of the ISO 14000 series

| Class and its item | Importance score (1–10) | Suggestion |
|---|---|---|
| **1  Governmental regulations**<br>1.1 Governmental administrative requirement on adopting the ISO 14000 series<br>1.2 Governmental encouragement on financial subsidies, e.g. tax deduction/return<br>1.3 Governmental encouragement on non-financial allowance<br>1.4 Pressure from the government | $\overline{C}_1$ | $Sug_1$ |
| **2  Technology conditions**<br>2.1 Reliable consultant companies on tutorship of adoption of the ISO 14000 series<br>2.2 Multifarious documental operation process of the ISO 14000 series<br>2.3 Destitute of applicability of the ISO 14000 series in construction enterprises<br>2.4 Suitable technology and material for environmental protection | $\overline{C}_2$ | $Sug_2$ |
| **3  Competitive pressures**<br>3.1 Competitive pressure from domestic construction industry<br>3.2 Competitive pressure from international construction industry<br>3.3 Pressure from the competitors inside construction industry<br>3.4 No competitors implemented the ISO 14000 series first inside construction industry<br>3.5 Pressure from the clients | $\overline{C}_3$ | $Sug_3$ |
| **4  Cooperative attitude**<br>4.1 Internal initiative consciousness on implementation of EMS<br>4.2 Correspondence and cooperation of design and construction<br>4.3 Employees' attitude towards cooperation on implementing the ISO 14000 series<br>4.4 Administrators' attitude towards cooperation on implementing the ISO 14000 series<br>4.5 Subcontractors' attitude towards cooperation on implementing the ISO 14000 series<br>4.6 Suppliers' attitude towards cooperation on implementing the ISO 14000 series | $\overline{C}_4$ | $Sug_4$ |

*Table B.4 (Continued)*

| Class and its item | Importance score (1–10) | Suggestion |
|---|---|---|
| **5  Cost–benefit efficiency** | $\overline{C}_5$ | $Sug_5$ |
| 5.1 Cost of implementation of ISO 14001 EMS | | |
| 5.2 Cost of ISO 14001 EMS assessment, certification, and maintenance | | |
| 5.3 Additional cost of human resource on adopting and implementing the ISO 14000 series | | |
| 5.4 Cost of ISO 14001 registration | | |
| 5.5 Additional cost of implementation of ISO 14001 EMS | | |
| 5.6 Impacts and additional expense of construction on interruption and adjustment | | |
| 5.7 High expense on implementation | | |
| $A_{ISO\ 14k}$ | | |

where $S_i$ represents the average importance score of class $i$ ($i = 1$ to 5), $\overline{C}_i$ represents the average value of class $i$, which is calculated from values given by accepters in the questionnaire survey, and the $Sug_i$ represents suggestions to be provided to the class $i$ about how one can improve performance on each item so as to achieve the requirements of the ISO 14000 series. The suggestions are developed from five useful findings from the analysis of the survey results:

- Contractors who give higher score to Class 1 have less intention to accept ISO 14000;
- Contractors who give higher score to Class 2 have more intention to accept ISO 14000;
- Contractors who give higher score to Class 3 have more intention to accept ISO 14000;
- Contractors who give higher score to Class 4 have more intention to accept ISO 14000;
- Contractors who give higher score to Class 5 have less intention to accept ISO 14000.

The model is implemented in a $45 \times 3$ spreadsheet of Microsoft Excel®, where the five classes and their associated items are listed in a checklist, as shown in Table B.4. Every item of the classes needs to be graded by users using the importance score ranging from 1 to 10, where 1 represents minimal importance and 10 represents maximal importance. When all items of the classes are valued, the spreadsheet generates average grade scores of each class and $A_{ISO\ 14k}$ is then calculated. If the value of $A_{ISO\ 14k}$ is 0, then suggestions ($Sug_i$) are provided to users on how to improve performance.

# Sample waste exchange websites

| Abbreviation | Website details |
| --- | --- |
| BCC | BuildFind Construction Classfieds <http://classifieds.buildfind.com/> (accessed in 2000–2003) |
| BRS | BuildingResources <http://www.buildingresources.org/> |
| BWE | Beyond Waste <http://www.sonic.net/~precycle/> |
| CDE | C&D Material Exchange <http://www.info.gov.hk/epd/misc/cdm/en_exchange1.html> |
| CMX | California Materials Exchange <http://www.ciwmb.ca.gov/CalMAX/> |
| CWE | The Commercial Waste Exchange <http://www.wastechange.com/> |
| HHM | Happy Harry's Used Building Materials <http://www.happyharry.com/hhub.htm> |
| HME | Hawaii Materials Exchange <http://www.maui.net/~mrghimex/himex1.html> |
| IDS | Industry Deals <http://www.industrydeals.com> (accessed in 2000–2003) |
| IME | Indiana Materials Exchange <http://www.state.in.us/idem/imex/> |
| IMX | Industrial Materials Exchange <http://www.metrokc.gov/hazwaste/imex/> |
| IWE | Illawarra Waste Exchange <http://www.globalpresence.com.au/waste_exchange/> |
| IWN | Industrial Waste Exchange Network <http://www.environ.wa.gov.au/iwe/> |
| IWX | Integrated Waste Exchange <http://www.capetown.gov.za/apps/iwe/default.asp> (accessed in 2000–2003) |
| KME | Kentucky Industrial Materials Exchange <http://www.kppc.org/kime/index.html> (accessed in 2000–2003) |
| LLR | Clubrecycle/Letsrecycle <http://www.letsrecycle.com/index.jsp> |
| MEX | Materials Exchange <http://www.cheltweb.com/wow/wexhome.htm> |
| MIE | Materials Information Exchange <http://cig.bre.co.uk/connet/mie/> |

(Continued)

| Abbreviation | Website details |
| --- | --- |
| MME | Minnesota Materials Exchange <http://www.mnexchange.org/> |
| MNY | Western New York Materials Exchange <http://recycle.net/recycle/exch/mat-ex/index.html> |
| NEE | New England Materials Exchange <http://www.wastecapnh.org/nemex/> (accessed in 2002–2003) |
| NHE | New Hampshire Materials Exchange <http://www.wastecapnh.org/nhme.htm> (accessed in 2002–2003) |
| NSE | Nova Scotia Material Exchange <http://www.clean.ns.ca/materials_exchange/wxh.htm> (accessed in 2000–2003) |
| NYE | New York Wa$te Match <http://www.wastematch.org/> |
| REN | Resource Exchange Network for Eliminating Waste <http://www.tnrcc.state.tx.us/exec/oppr/renew/renew.html> |
| RME | Reusable Building Materials Exchange <http://www.rbme.com/> |
| RSW | Recycler's World <http://www.recycle.net/build/index.html> (accessed in 2000–2003) |
| SCE | Sonoma County's Materials exchange <http://www.recyclenow.org/sonomax/> |
| SEE | Southern New England Materials Exchange <http://www.rirrc.org/materials.shtml> (accessed in 2000–2003) |
| SME | Southeast Minnesota Recyclers' Exchange <http://www.semrex.org/> |
| SWC | Solid Waste.com <http://www.solidwaste.com/content/homepage/default.asp> |
| TME | Tennessee Materials Exchange <http://www.cis.utk.edu/tme_titl.htm> (accessed in 2000–2003) |
| VCE | Ventura county Materials Exchange <http://www.rain.org/~swmd/vcmax/> (accessed in 2000–2003) |
| WME | Waste Management Commodities Exchange <http://commodities.wm.com/wmx/exchange.nsf> |
| WRA | The Waste & Resources Action Programme <http://www.wrap.org.uk/> |
| WUK | Waste Exchange UK <http://www.wasteexchangeuk.com/Template.htm> (accessed in 2000–2003) |

# Webfill function menu

The Webfill e-commerce system provides its members the following function menu for selection:

Account management

- Member's profile update
- Member's exchange record check
- Request buying/selling/bidding

☐ Buyers (Contractor/Managers/Manufacturers/Recycler/Disposers)

    ☐ Current information for buyers
    ☐ Search

        ☐ By type/category of *WasteSpec* (Kincaid, Walker and Flynn 1995)
        ☐ By brand of recovered materials
        ☐ By status (fixed price, bidding price, rent)
        ☐ By date

    ☐ Bid

        ☐ By type/category
        ☐ By bidding code of current bidding item on buyers' bulletin board

    ☐ Request

        ☐ By type/category
        ☐ By code of current item on buyers' bulletin board

☐ Sellers (Contractor/Managers/Manufacturers/Recyclers/Disposers)

    ☐ Current information for sellers
    ☐ Search

        ☐ By type/category
        ☐ By brand of recovered materials

- ☐ By status (fixed price, bidding price, rent)
- ☐ By date

☐ Bid

- ☐ By type/category
- ☐ By bidding code of current bidding item on sellers' bulletin board

☐ Request

- ☐ By type/category
- ☐ By code of current item on sellers' bulletin board

☐ Transporters

☐ Current information for transporters
☐ Search

- ☐ By type/category
- ☐ By master of goods (C&D waste or recovered materials)
- ☐ By location
- ☐ By date

☐ Bid

- ☐ By type/category
- ☐ By bidding code of current bidding item on sellers' bulletin board

☐ Request

- ☐ By type/category
- ☐ By code of current item on sellers' bulletin board

# Glossary

*Construction Pollution Index (CPI)*: It is a method to quantitatively measure the amount of environmental pollution and hazards generated from individual processes and the whole project during construction; it can be utilized by indicating the potential level of accumulated environmental pollution and hazards generated from a construction site, and by reducing or mitigating pollution level during construction planning stage.

*Environmental Impacts Assessment (EIA)*: It is a process to identify, predict, evaluate, and mitigate the biophysical, social, and other relevant environmental effects of development proposals or projects prior to major decisions being taken and commitments made.

*Env.Plan*: It is a multicriteria decision-making model for evaluating construction plan alternatives based on analytic network process (ANP) theory and experts' knowledge. The CPI and the env.Plan are two essential tools for preventing potential adverse environmental impacts at pre-construction stage.

*E+*: It is an integrative methodology for effective, efficient, and economical EM in construction projects, in which an EMS-based dynamic EIA process is applied inside a knowledge-driven decision-support system for active knowledge capture and re-use focusing on environmental-conscious construction management.

*Incentive Reward Program (IRP)*: It is a financial incentive program (FIR), which can be utilized as an on-site material and equipment management system to control and reduce construction waste. Information systems using bar-code technology or radio frequency identification (RFID) technology can facilitate its implementation.

*Webfill*: It is an online waste exchange approach or an e-commerce business plan, which is developed based on e-commerce theory and the trip-ticket system being used for waste disposal management in Hong Kong. It can be utilized to reduce the final amount of construction and demolition (C&D) waste to be land-filled.

# References

Abdelhamid, T.S., and Everett, J.G. (1999). Physiological demands of concrete slab placing and finishing work. *Journal of Construction Engineering and Management*, ASCE, 125(1), 47–52.

Abdullah, A., and Anumba, C.J. (2002a). Decision criteria for the selection of demolition techniques. *Proceeding of the Second International Postgraduate Research Conference in the Built and Human Environment*. University of Salford, 11–12 April 2002, edited by M. Sun, *et al.* Blackwell Publishers. 410–419.

Abdullah, A., and Anumba, C.J. (2002b). Decision model for the selection of demolition techniques. *Proceeding of the International Conference on Advances in Building Technology*. The Hong Kong Polytechnic University, 4–6 December 2002. Elsevier Science Limited. 1671–1681.

Abudayyeh, O., Sawhney, A., El-Bibany, H., and Buchanan, D. (1998). Concrete bridge: Demolition methods and equipment. *Journal of Bridge Engineering*, 3(3), 117–125.

Adams, T.M., Malaikrisanachalee, S., Blazquez, C., and Vonderohe, A. (2000). GIS-based automated oversize/overweight permit processing. In R. Fruchter (editor), F. Peña-Mora, W.M.K. Roddis (Ed.), *Computing in Civil and Building Engineering* (Proceedings of the Eighth International Conference held in Stanford, California, 14–16 August 2000), ASCE, Reston, 209–216.

Adams, T.M., Vonderohe, A.P., Russell, J.S., and Clapp, J.L. (1992). Integrating facility delivery through spatial information. *Journal of Urban Planning and Development*, ASCE, 118(1), 13–23.

Adeli, H., and Karim, A. (2001). *Construction Scheduling, Cost Optimization and Management: A New Model Based on Neurocomputing and Object Technologies*. Spon Press, London and New York.

Ammenberg, J., Borjesson, B., and Hjelm, O. (2000). Joint EMS and group certification: A cost-effective route for SMEs to achieve ISO 14001. In *ISO 14001: Case Studies and Practical Experiences*, edited by H. Ruth. Greenleaf Publishing Ltd, Sheffield, UK. 58–66.

Anumba, C.J., Abdullah, A., and Fesseha, T. (2003). Selection of demolition techniques: A case study of the Warren Farm bridge. *Structural Survey*. Emerald, MCB UP Limited, 21(1), 36–48.

Arnfalk, P. (1999). *Information Technology in Pollution Prevention: Teleconferencing and Telework used as Tools in the Reduction of Work Related Travel*, IIIEE Dissertations, the International Institute for Industrial Environmental Economics at Lund University, Sweden.

Austin, T. (1991). Building green. *Civil Engineering*, ASCE, 61(8), 52–54.

Azani, C.H. (1999). An integrative methodology for the strategic management of advanced integrated manufacturing systems, in *Advanced Manufacturing Systems: Strategic Management and Implementation*, edited by J. Sarkis and H.R. Parsaei. Gordon and Breach Science Publishers, Australia, 21–41.

Bakken, J.D., and Avey, C.M. (1992). Integration of AM/FM/GIS with MODELING/DESIGN on large utility PC network. In B.J. Goodno, J.R. Wright (Editors), *Computing in Civil Engineering and Geographic Information Systems Symposium* (Proceedings of the Eighth Technical Council on Computer Practices Conference held in conjunction with A/E/C Systems '92), ASCE, New York, 703–711.

Baldwin, A.N., Thorpe, A., and Alkaabi, J.A. (1994). Improved materials management through bar-coding: Results and implications from a feasibility study. *Proceedings of Institute of Civil Engineers: Civil Engineering*, 102(6), 156–162.

Bell, L.C., and McCullouch, B.G. (1988). Bar code application in construction. *Journal of Construction Engineering and Management*, ASCE, 114(2), 263–278.

Bello, D., Virji, M.A., Kalil, A.J., and Woskie, S.R. (2002). Quantification of respirable, thoracic, and inhalable quartz exposures by FT-IR in Personal Impactor Samples from Construction Sites. *Applied Occupational and Environmental Hygiene*, Taylor and Francis Ltd, 17(8), 580–590.

Berning, P.W., and Diveley-Coyne, S. (2000). E-Commerce and the construction industry: The revolution is here. *Industry Reports Newsletters*. Thelen Reid and Priest LLP, New York. <http://www.constructionweblinks.com/Resources/Industry_ Reports__ Newsletters/Oct_2_2000/e-commerce.htm> (20 August 2002).

Bernold, L.E. (1990a). Testing barcode technology in construction environment. *Journal of Construction Engineering and Management*, ASCE, 116(4), 643–655.

Bernold, L.E. (1990b). Barcode-driven equipment and materials tracking for construction. *Journal of Computing in Civil Engineering*, ASCE, 4(4), 381–395.

Bernold, L.E. (2002). Spatial Integration in construction. *Journal of Construction Engineering and Management*, ASCE, 128(5), 400–408.

Bernstein, C.S. (1983). Highway Projects – Can they be done in half the time? *Civil Engineering*, ASCE, 53(9), 50–54.

Blakey, L.H. (1990). Barcode: prescription for precision, performance, and productivity. *Journal of Construction Engineering and Management*, ASCE, 116(3), 468–479.

Boggess, G., and Abdul, M. (1997). *The Application of Genetic Algorithms to the Scheduling of Engineering Units*. A Report to the U. S. Army Corps of engineers Waterways Experiment Station Geotechnical Laboratory Mobility Systems Division. Computer Science Department, Mississippi State University. USA.

Bonforte, G.A., and Keeber, G. (1993). Tunnel repairs under traffic and community impacts. *Infrastructure Planning and Management* (consists of papers presented at two parallel conferences held in Denver, Colorado, 21–23 June 1993). ASCE, New York, 187–191.

Bossink, B.A.G., and Brouwers, H.J.H. (1996). Construction waste: Quantification and source evaluation. *Journal of Construction Engineering and Management*, ASCE, 122(1), 55–60.

Bossler, J.D. (2001). *Manual of Geospatial Science and Technology*, Taylor & Francis London and New York.

Brandon, T.L., and Stadler, R.A. (1991). Use of barcode technology to simplify sieve analysis data acquisition. *Geotechnical Engineering Congress 1991* (Volume 1): Proceedings of the Congress sponsored by the Geotechnical Engineering Division of the American Society of Civil Engineers. (Edited by F.G. McLean, D.A. Campbell and D.W. Harris) (Geotechnical Special Publication No. 27), ASCE, 556–561.

BSI (2000). *BS 6187:2000 – Code of Practice for Demolition*. British Standards Institution (BSI), British Standards Publishing Limited (BSPL). UK.

CACEB (2002). *The Directory of ISO 14001 Certified Companies (as at 31 December 2001)*. China Registration Committee for Environmental Management System Certification Bodies (CACEB), China. <http://www.naceca.org/> (26 December 2002).

Carberry, E. (1996). Assessing ESOPs. *Journal of Management in Engineering*, ASCE, 12(5), 17–19.

Carper, K.A. (1990). Environmental permitting a major expressway facility in Florida. *Optimizing the Resources for Water Management: Proceedings of the 17th Annual National Conference*, Water Resources Planning and Management Division, ASCE, Fort Worth, Texas, USA, 144–148.

CCEMS (China Center for EMS) (2001). *The Directory of ISO 14001 Certified Companies (as at 23 November, 2001)*. <http://www.ccems.com.cn/news/news-index.html> (23 November 2001).

CEC (2001). *The Directory of ISO 14001 Certified Companies by CEC (as at 22 November, 2001)*. Certification Center of Environmental Management System (CEC), Chinese Research Academy of Environmental Sciences, China. <http://www.chinaiso14000-series.com/roster1.htm> (22 November 2001).

CED (2002). *Management of Public Filling Facilities and Dumping Licences*. Civil Engineering Department (CED), Hong Kong ASR Government, Hong Kong. <http://www.info.gov.hk/ced/eng/services/licences/licence.htm> (2 August 2002).

CEIN (2001a). *Main Indicators on Construction Enterprises*. <http://www.cein.gov.cn/home/hytj/jzyhytj/1998/14-1.doc> (22 November 2001).

CEIN (2002). *Main Indicators on Construction Enterprises*. China Engineering Information Net (CEIN), China. <http:// www.cein.gov.cn/> (26 December 2002).

Chan, W.T., Chua, D.K.H., and Kannan, G. (1996). Construction resource scheduling with genetic algorithms. *Journal of Construction Engineering and Management*, ASCE, 122(2), 125–132.

Chang, T.C., William, C., and Crandall, K.C. (1990). Network resource allocation with support of a fuzzy expert system. *Journal of Construction Engineering and Management*, ASCE, 116(2), 239–259.

Chen, Z. (2003). *An Integrated Analytical Approach to Environmental Management in Construction*. Ph.D. Dissertation. Department of Building and Real Estate, Hong Kong Polytechnic University. Hong Kong. ProQuest, USA. UMI Number: AAT 3107430. <http://wwwlib.umi.com/dissertations/ preview/3107430>

Chen, Z., and Li, H. (2003). An analytic network process model for demolition planning. *Proceedings of the CIB Student Chapters International Symposium on Innovation in Construction and Real Estate*. September 2003, Hong Kong Polytechnic University, Hong Kong.

Chen, Z., and Li, H. (2005). A knowledge-driven management approach to environmental-conscious construction. *International Journal of Construction Innovation*, Hodder Arnold, 5(1), 27–39.

Chen, Z., Li, H., and Wong, C.T.C. (2000). Environmental management of urban construction projects in China. *Journal of Construction Engineering and Management*, ASCE, 126(4), 320–324.

Chen, Z., Li, H., and Wong, C.T.C. (2002a). An application of bar-code system for reducing construction wastes. *Automation in Construction*, 11(5), 521–533.

Chen, Z., Li, H., Wong, C.T.C., and Love, P.E.D. (2002c). Integrating construction pollution control with construction schedule: An experimental approach. *Environmental Management and Health*, 13(2), 142–151.

Chen, Z., Li, H., and Wong, C.T.C. (2003a). Webfill before landfill: An e-commerce model for waste exchange in Hong Kong. *Journal of Construction Innovation*, 3(1), 27–43.

Chen, Z., Li, H., and Wong, C.T.C. (2003b). Environmental priority evaluation for construction planning. *Proceedings of the 2nd International Conference on Innovation in Architecture, Engineering and Construction (AEC)*. June 2003. Loughborough University, UK.

Chen, Z., Li, H., and Wong, C.T.C. (2003c). E+: An integrative methodology for dynamic EIA in construction. *Proceedings of the 3rd International Post-Graduate Research Conference*. April 2003. Salford University, Lisbon, Portugal.

Chen, Z., Li, H., and Wong, C.T.C. (2003d). An integrative methodology for dynamic EM in construction. *Proceedings of the ARCOM Doctoral Workshop*. 18 June 2003. Glasgow Caledonian University, Glasgow, UK.

Chen, Z., Li, H., and Hong, J. (2004a). An integrative methodology for environmental management in construction. *Automation in Construction*, 13(5), 621–628.

Chen, Z., Li, H., Shen, Q.P., and Xu, W. (2004b). An empirical model for decision-making on ISO 14000. *Construction Management and Economics*, 22(1), 55–73.

Chen, Z., Li, H., and Wong, C.T.C. (2005). EnvironalPlanning: An analytic network process model for environmentally conscious construction planning. *Journal of Construction Engineering and Management*, ASCE, 131(1), 92–101.

Cheng, M.Y., and O'Connor, J.T. (1996). ArcSite: Enhanced GIS for construction site layout. *Journal of Construction Engineering and Management*, ASCE, 122(4), 329–336.

Cheng, M.Y., and Yang, S.C. (2001). GIS-based cost estimates integrating with material layout planning. *Journal of Construction Engineering and Management*, ASCE, 127(4), 291–299.

Cheung, C.M., Wong, K.W., Poon, C.S., Fan, C.N., and Cheung, A.C. (1993). *Reduction of Construction Waste: Final Report*. Department of BRE and CSE of the Hong Kong Polytechnic University and the Hong Kong Construction Association Ltd, Hong Kong.

China Environment Daily (16 December 2002). Bureau confirms environmental supervision in pilot national construction projects. *China Environment Daily*, China, p. A4. <http://search.envir.com.cn/info/2002/12/1216921.htm> (26 December 2002).

China EPB (2000). *Official Report on the State of the Environment in China 1999*. Environmental Protection Bureau (EPB), China. <http://www.zhb.gov.cn/bulletin/soechina99/index.htm> (22 November 2001).

China EPB (2002). *Official Report on the State of the Environment in China*. Environmental Protection Bureau (EPB), China. <http://www.zhb.gov.cn/bulletin/country_intro.php3> (26 December 2002).

China NBS (1998). *Statistical Yearbook of China 1997*. National Bureau of Statistics (NBS), China. <http://www.stats.gov.cn/sjjw/ndsj/information/nj97/ml97.htm> (22 November 2001).

China NBS (2000). *Statistical Yearbook of China 1999*. National Bureau of Statistics (NBS), China. <http://www.stats.gov.cn/yearbook/indexC.htm> (22 November 2001).

China NBS (2001). *Statistical Yearbook of China 2000*. National Bureau of Statistics (NBS), China. <http://www.stats.gov.cn/ sjjw/ndsj/zgnj/mulu.html> (22 November 2001).

Chua, D.K.H., Chan, W.T., and Kannan, G. (1996). Scheduling with Co-Evolving Resource Availability Profiles. *Civil Engineering System*, 13(6), 311–329.

CIOB (1997). *Code of Estimating Practice* (6th edition). The Chartered Institute of Building (CIOB), Longman, England.

CIRIA (1993). *Environmental Issues in Construction – A Review of Issues and Initiatives Relevant to the Building, Construction and Relevant Industries. Volume 2 – Technical Review*. Publication SP94, Construction Industry Research and Information Association (CIRIA), Thomas Telford, London.

CIRIA (1994a). *Environmental Assessment*. Publication SP96, Construction Industry Research and Information Association (CIRIA), Thomas Telford, London.

CIRIA (1994b). *Environmental Handbook for Building and Civil Engineering Projects: Checklists, Obligations, Good Practice and Sources of Information*. Publication SP97 and 98, Construction Industry Research and Information Association (CIRIA), Thomas Telford, London.

CIRIA (1995). *A Clients Guide to Greener Construction: A Guide to Help Clients Address the Environmental Issues to be Faced on Building and Civil Engineering Projects*. Publication 120, Construction Industry Research and Information Association (CIRIA), London.

CIRIA (1999). *Environmental Issues in Construction – Research Campaign. Executive Summary*. Publication PR74, Construction Industry Research and Information Association (CIRIA), London.

Clough, R.H., and Antonio, N. (1996). *Environmental Management in Construction: Model Forms to Assist Implementation*. The Chartered Institute of Building (CIOB), UK.

CMC (2000). *Main Applied Construction Technologies in the 10th Five-year Plan*. Official Document No. (2000)286, Technology Department, China Ministry of Construction (CMC), China. <http://www.cein.gov.cn/ show.asp?rec_no=225> (22 November 2001).

CMX (2000). California Materials Exchange (CalMAX). <http://www.ciwmb.ca.gov/ CalMAX/> (20 August 2002) (Reached first in 2000).

Coventry, S., and Woolveridge, C. (1999). *Environmental Good Practice on Site*. Construction Industry Research and Information Association (CIRIA), London.

Coventry, S., Woolveridge, C., and Patel, V. (1999). *Waste Minimisation and Recycling in Construction – Boardroom Handbook*. Special Publication 135, Construction Industry Research and Information Association (CIRIA), London.

CPSC (1998). *Environmental Management Systems – Guidelines*. Construction Policy Steering Committee (CPSC), NSW Government, Australia. <http://www.cpsc.nsw.gov.au/environment/> (22 November 2001).

CPSC (2001). Construction Industry CEO Survey: NSW Construction Industry Survey of Industry Leaders. Construction Policy Steering Committee (CPSC), NSW Government, Australia. <http://www.cpsc.nsw.gov.au/docs/ strategic-info/CEO-Survey-Fax-2001.pdf> (22 November 2001).

Darnall, N. (2001). Why some firms mandate ISO 14001 certification while others encourage it. Paper for presentation at the Twenty-Third Annual Research Conference for the Association for Public Policy Analysis and Management Fall Conference: "Public Policy Analysis and Public Policy: Making the Connection", 1–3 November Washington Monarch Hotel, Washington, DC.

Davis, E.W., and Patterson, J.H. (1975). A comparison of heuristic and optimum solutions in resource-constrained project scheduling. *Management Science*, The Institute of Management Sciences, 21(8), 944–955.

Davis, R., Shrobe, H., and Szolovits, P. (1993). What is a Knowledge Representation? *AI Magazine*, 14(1), 17–33.

DeMocker, J. (1999). Building in efficiency. *InternetWeek*. CMP Media LLC. <http://www.internetweek.com/transform/transform112299-1.htm> (26 June 2005).

Dey, P.K., Tabucanon, M.T., and Ogunlana, S.O. (1996). Petroleum pipeline construction planning: A conceptual framework. *International Journal of Project Management*, Elsevier Science Ltd and IPMA, 14(4), 231–240.

Dodds, P.J., and Sternberger, R.S. (1992). The evolution of an environmental monitor. *Civil Engineering*, ASCE, 62(6), 56–58.

Dohrenwend, R.E. (1973). Environmental management during power transmission line construction: Operational considerations. *Annual Meeting Proceedings of the Colloq Biotic Manage Along Power Transmit Rights of Way*, American Institute of Biology Science, Amherst, Mass, USA. NY Bot Gard, Cary Arbor, Millbrook, 58–77.

Echeverry, D. (1996). Adaptation of barcode technology for construction project control. *Computing in Civil Engineering: Proceedings of International Computing Congress in Civil Engineering* (3rd), edited by J. Vanegas and P. Chinowsky, ASCE, Reston, USA, 1034–1040.

Echeverry, D., and Beltran, A. (1997). Barcode control of construction field personnel and construction materials. *Computing in Civil Engineering: Proceedings of International Computing Congress in Civil Engineering* (4th), edited by T.M. Adams, ASCE, Reston, USA, 341–347.

Echeverry, D., Guerra, C.A., and Beltran, A. (1998). A regional attempt to implement barcode control in construction project. In *Computing in Civil Engineering: Proceedings of International Computing Congress in Civil Engineering* (5th), edited by K.C.P. Wang, T. Adams, M.L. Maher and A. Songer, ASCE, Reston, USA, 450–453.

Emery, J.J. (1974). Use of mining and metallurgical waste in construction. *Proceedings of International Symposium on Miner and the Environment*, ASME, Institute of Mine and Metall, London, 261–272.

Enkawa, T., and Schvaneveldt, S. (2001). Just-in-Time, Lean Production, and Complementary Paradigms. In *Handbook of Industrial Engineering: Technology and Operations Management* (3rd edition) edited by G. Salrendy, John Wiley & Sons, Inc., New York. ISBN: 0-471-33057-4.

U.S.EPA (2000). *EPA Region 9 Solid Waste Program: Construction and Demolition (C&D) Debris*. Environmental Protection Agency (EPA), USA. Available at http://www.epa.gov/region09/waste/solid/debris.htm

U.S.EPA (2004a). *FIELDS methods*. Environmental Protection Agency (EPA), USA. <http://www.epa.gov/region5fields/ htm/methods.htm> (9 August 2004).

U.S.EPA (2004b). *GPS – Global Positioning System*. Environmental Protection Agency (EPA), USA. <http://www.epa.gov/region5fields/htm/methods/gps/> (9 August 2004).

U.S.EPA (2004c). *What is GIS (Geography Information Systems)?* Environmental Protection Agency (EPA), USA. <http://www.epa.gov/region5fields/htm/methods/gis/> (9 August 2004).

HKEPD (1997a). *Environment Hong Kong 1997.* Environmental Protection Department (EPD), Hong Kong. <http://www.epd.gov.hk/epd/english/resources_pub/publications/pub_reports_ap.html> (26 June 2005).

HKEPD (1998a). *Environment Hong Kong 1998.* Environmental Protection Department (EPD), Hong Kong. <http://www.epd.gov.hk/epd/english/resources_pub/publications/pub_reports_ap.html> (26 June 2005).

HKEPD (1998b). *Monitoring of Solid Waste in Hong Kong 1997.* Environmental Protection Department (EPD), Hong Kong. <http://www.epd.gov.hk/epd/english/environmentinhk/waste/data/waste_mon_swinhk.html> (26 June 2005).

HKEPD (1998d). *Solid Waste Statistics 1998 Updates.* Environmental Protection Department (EPD), Hong Kong. <http://www.info.gov.hk/epd/E/pub/sw-rep/98update/Index.htm> (11 August 2004).

HKEPD (1999a). *Environment Hong Kong 1999.* Environmental Protection Department (EPD), Hong Kong. <http://www.epd.gov.hk/epd/english/resources_pub/publications/pub_reports_ap.html> (26 June 2005).

HKEPD (1999b). *Monitoring of Solid Waste in Hong Kong 1998.* Environmental Protection Department (EPD), Hong Kong. <http://www.epd.gov.hk/epd/english/environmentinhk/waste/data/waste_mon_swinhk.html> (26 June 2005).

HKEPD (1999c). *Environmental Protection in Hong Kong.* Environmental Protection Department (EPD), Hong Kong.

HKEPD (1999d). *Trip-ticket System for Disposal of Construction and Demolition Material.* Works Bureau Technical Circular No. 5/99. Environmental Protection Department (EPD), Hong Kong.

HKEPD (2000a). *Environment Hong Kong 2000.* Environmental Protection Department (EPD), Hong Kong. <http://www.epd.gov.hk/epd/english/resources_pub/publications/pub_reports_ap.html> (26 June 2005).

HKEPD (2000b). *Monitoring of Solid Waste in Hong Kong 1999.* Environmental Protection Department (EPD), Hong Kong. <http://www.epd.gov.hk/epd/english/environmentinhk/waste/data/waste_mon_swinhk.html> (26 June 2005).

HKEPD (2000c). *Environmental Impact Assessment Ordinance.* Chapter 499, Section 10. Environmental Protection Department (EPD), Hong Kong.

HKEPD (2001a). *Environment Hong Kong 2001.* Environmental Protection Department (EPD), Hong Kong. <http://www.epd.gov.hk/epd/english/resources_pub/publications/pub_reports_ap.html> (26 June 2005).

HKEPD (2001b). *Monitoring of Solid Waste in Hong Kong 2000.* Environmental Protection Department (EPD), Hong Kong. <http://www.epd.gov.hk/epd/english/environmentinhk/waste/data/waste_mon_swinhk.html> (26 June 2005).

HKEPD (2002a). *Environment Hong Kong 2002.* Environmental Protection Department (EPD), Hong Kong. <http://www.epd.gov.hk/epd/english/resources_pub/publications/pub_reports_ap.html> (26 June 2005).

HKEPD (2002b). *Monitoring of Solid Waste in Hong Kong 2001.* Environmental Protection Department (EPD), Hong Kong. <http://www.epd.gov.hk/epd/english/environmentinhk/waste/data/waste_mon_swinhk.html> (26 June 2005).

HKEPD (2002c). *C&D Material Exchange.* Environmental Protection Department (EPD), Hong Kong. <http://www.info.gov.hk/epd/misc/cdm/ en_exchange.html> (20 August 2002).

HKEPD (2003a). *Environment Hong Kong 2003*. Environmental Protection Department (EPD), Hong Kong. <http://www.epd.gov.hk/epd/english/resources_pub/publications/pub_reports_ap.html> (26 June 2005).

HKEPD (2003b). *Monitoring of Solid Waste in Hong Kong 2002*. Environmental Protection Department (EPD), Hong Kong. <http://www.epd.gov.hk/epd/english/environmentinhk/ waste/data/waste_mon_swinhk.html> (26 June 2005).

HKEPD (2004a). *Environment Hong Kong 2004*. Environmental Protection Department (EPD), Hong Kong. <http://www.epd.gov.hk/epd/english/resources_pub/ publications/pub_reports_ap.html> (26 June 2005).

HKEPD (2004b). *Monitoring of Solid Waste in Hong Kong 2003*. Environmental Protection Department (EPD), Hong Kong. <http://www.epd.gov.hk/epd/english/environmentinhk/ waste/data/waste_mon_swinhk.html> (26 June 2005).

HKEPD (2005a). *Environment Hong Kong 2005*. Environmental Protection Department (EPD), Hong Kong. <http://www.epd.gov.hk/epd/english/ resources_pub/publications/pub_reports_ap.html> (26 June 2005).

HKEPD (2005b). *Monitoring of Solid Waste in Hong Kong 2004*. Environmental Protection Department (EPD), Hong Kong. <http://www.epd.gov.hk/epd/english/environmentinhk/waste/data/waste_mon_swinhk.html> (26 June 2005).

Escanciano, C., Fernandez, E., and Vazquez, C. (2002). Linking the firm's technological status and ISO 9000 certification: Results of an empirical research. *Technovation*, 22(8), 509–515.

ESRI (2004). *ArcGIS*. <http://www.esri.com/software/arcgis/about/overview.html>

European Commission (1999). *Integrating Environment Concerns into Development and Economic Cooperation*. Draft version 1.0, Brussels. Retrieved from European Environment Agency. <http://glossary.eea.eu.int/EEAGlossary/E/environmental_impact_assessment> (22 November 2001).

Farid, F., and Manoharan, S. (1996). Comparative analysis of resource-allocation capabilities of project management software packages. *Project Management Journal*, 35–44 June.

Fesseha, T. (1999). *Criteria for Selection of Demolition Techniques*. MSc Thesis, Loughborough University, Loughborough, UK.

FIDIC (1998). *Guide to ISO 14001 Certification/Registration* (Test edition). International Federation of Consulting Engineers (FIDIC), Switzerland.

Field, A. (2000). *Discovering Statistics Using SPSS for Windows*. SAGE Publications, London.

Fishbein, B.K. (1998). *Building for the Future: Strategies to Reduce Construction and Demolition Waste in Municipal Projects*. INFORM, Inc. <http://www.informinc.org/cdreport.html> (15 May 2002).

Frics, J.W. (1996). *Estimating for Building and Civil Engineering Works* (9th edition). Butterworth Heinemann Ltd, Oxford.

Gambatese, J.A., and James, D.E. (2001). Dust suppression using truck-mounted water spray system. *Journal of Construction Engineering and Management*, ASCE, 127(1), 53–59.

Gavilan, R.M., and Bernold, L.E. (1994). Source evaluation of solid waste in building construction. *Journal of Construction Engineering and Management*, ASCE, 120(3), 536–552.

Gidley, J.S., and Sack, W.A. (1984). Environmental aspects of waste utilization in construction. *Journal of Environmental Engineering*, ASCE, 110(6), 1117–1133.

Google (2005). Google Directory: Science > Software > Simulation. <http://directory.google.com/Top/Science/Software/Simulation/> (26 June 2005).

Griffith, A. (1994). *Environmental Management in Construction*. MacMillan Press Ltd, London.

Griffith, A., Stephenson, P., and Watson, P. (2000). *Management System for Construction*. Pearson Education Inc., New York and Englemere Ltd, England.

Grigg, N.S., Criswell, W.E., Fontane, D.G., and Siller, T.J. (2001). *Civil Engineering Practice in the Twenty-First Century: Knowledge and Skills for Design and Management*. ASCE Press, Reston, USA, 264 pp.

Griss, M., and Letsinger, R. (2000). Games at work – agent-mediated e-commerce simulation. In *HP Labs 2000 Technical Reports*. HP Labs, Hewlett-Packard Development Company, L.P. <http://www.hpl.hp.com/ techreports/2000/HPL-2000-52.html> (26 June 2005).

Grobler, F., Kannan, G., Subick, C., and Kargupta, H. (1995). Optimization of uncertain resource plans with GA. *Computing in Civil Engineering (1995)* (Proceedings of the Second Congress held in conjunction with A/E/C Systems '95 held in Atlanta, Georgia, 5–8 June 1995), ASCE, 1643–1650.

Guthrie, P., and Mallett, H. (1995). *Waste Minimization and Recycling in Construction: A Review*. Publication 122, CIRIA, London.

Hammad, A., Itoh, Y., and Nishido, T. (1993). Bridge planning using GIS and expert system approach, *Journal of Computing in Civil Engineering*, ASCE, 7(3), 278–295.

Hampton, T. (2004). 10 electronic technologies that changed construction (6/21/2004 Issue), ENR, The McGraw-Hill Companies, Inc. <http://enr.construction.com/features/ technologyEconst/archives/040621.asp> (26 June 2005).

Harris, R. (1978). *Resource and Arrow Networking Techniques for Construction*. Wiley, New York.

HB (2000). *Hong Kong Fact Sheet – Housing (1999)*. Housing, Planning and Lands Bureau (HB), Hong Kong. <http://www.info.gov.hk/hkfacts/ housing.pdf> (26 June 2005).

Hegazy, T. (1999). Optimization of construction time-cost trade-off analysis using genetic algorithms. *Canada Journal of Civil Engineering*, NRC Canada, 26, 685–697.

Hegazy, T. (1999). Optimization of resource allocation and leveling using Genetic Algorithms. *Journal of Construction Engineering and Management*, ASCE, 125(3), 167–175.

Henderson, R.D. (1970). *Air Pollution and Construction Equipment*. SAE Earthmoving Industry Conference, Peoria, IL, 14–15, April. Paper 700551. 6.

Hendrickson, C., and Au, T. (2000). *Project Management for Construction: Fundamental Concepts for Owners, Engineers, Architects and Builders* (2nd edition). <http:// www.ce.cmu.edu/pmbook/>(01/01/2006). 1st Edition printed by Prentice Hall, 1989.

Hendrickson, C., and Horvath, A. (2000). Resource use and environmental emissions of U.S. construction sectors. *Journal of Construction Engineering & Management*, ASCE, 126(1), 38–44.

Henningson, J.C. (1978). Environmental management during construction. *Journal of the Construction Division*. ASCE, 104(4), 479–485.

Hinckley, J.M. (1986). Reviewing for potential failure. *Civil Engineering*, ASCE, 56(7), 60–62.

HKEPD (2002). *The Directory of ISO 14001 Certified Companies in Hong Kong* (as at 2 April 2002). Hong Kong Environmental Protection Department (HKEPD), Hong Kong. <http://www.info.gov.hk/epd/english/how_help/tools_cem/ iso14001.html> (1 July 2002).

HKPC (1996). HKPC Updates Local Businesses on ISO 14000 series and Environmental Management. *HKPC Productivity News*. Hong Kong Productivity Council (HKPC), Hong Kong, 55–56, October.

HKPC (2000). Press Release: *HKPC Organizes Industry Support Scheme on ISO 14001 Environmental Management System for Small and Medium Enterprises.* 22 February. Hong Kong Productivity Council (HKPC), Hong Kong. <http://www.hkpc.org/hkpc/html/newc_press_unit.asp?unit=4> (22 November 2001).

Ho, L. (1997). *Human Resources Planning Strategies of the Hong Kong Construction Industry.* Thesis of MBA. Department of Management, The Hong Kong Polytechnic University, Hong Kong.

Horvath, A., and Hendrickson, C. (1998). Steel versus steel-reinforced concrete bridges: environmental assessment. *Journal of Infrastructure Systems*, ASCE, 4(3), 111–117.

IAIA (1997). *Principles of Environmental Impact Assessment Best Practice.* International Association for Impact Assessment (IAIA). <http://www.iaia.org/principles/> (26 December 2002).

IDC (2004). *Business Simulations – Stimulating Business.* IDC Interact Ltd, Bristol, UK. <www.idcinteract.com>

ILS (2003). *Strategic E-Commerce Simulation.* Innovative Learning Solutions, Inc. (ILS). <http://www.marketplace-simulation.com/products/strategic-e-commerce.php> (26 June 2005).

Islam, M.S.P.E., and Hashmi, S.E.Q.P.E. (1999). Geotechnical aspects of foundation design and construction for pipelines buried in liquefiable riverbed. *Optimizing Post-Earthquake Lifeline System Reliability: Proceedings of the 5th U.S. Conference on Lifeline Earthquake Engineering*, ASCE, Seattle, WA, USA, 389–400.

ISO (The International Organization for Standardization) (2001). *ISO Survey of ISO 9000 and ISO 14000 series Certificates.* (Tenth cycle: up to and including 31 December 2000) <http://www.iso.org/iso/en/CombinedQueryResult.CombinedQueryResult?queryString=Survey> (4 June 2002).

ISO (2002). *ISO 9000 and ISO 14000 Certifications Reach Record Levels in 2001.* The International Organization for Standardization (ISO). 19 July 2002. Ref.: 830. <http://www.iso.org/> (26 December 2002).

Issa, R.R.A. (1995). A pen-centric application for construction quality and Productivity Tracking. In J.P. Mohsen (Editor), *Computing in Civil Engineering* (Proceedings of the Second Congress held in conjunction with A/E/C Systems '95 held in Atlanta, Georgia, June 5–8, 1995), ASCE, New York, 1356–1359.

Jeljeli, M.N., and Russell, J.S. (1995). Coping with uncertainty in environmental construction: decision-analysis approach. *Journal of Construction Engineering and Management*, ASCE, 121(4), 370–380.

Jeljeli, M.N., Russell, J.S., Meyer, H.W.G., and Vonderohe, A.P. (1993). Potential applications of geographic information systems to construction industry. *Journal of Construction Engineering and Management*, ASCE, 119(1), 72–86.

Jones, K.H. (1973). Synthesis approach to determining research needs in civil engineering. *Journal of the Environmental Engineering Division*, ASCE, 99(4), 461–467.

Jones, N., and Klassen, R.D. (2001). Management of pollution prevention: Integrating environmental technologies in manufacturing. In *Greener Manufacturing and Operations: From Design to Delivery and Back.* Edited by J. Sarkis, J. Greenleaf Publishing Ltd. Sheffield, UK, 56–68.

Kasai, Y. (ed.) (1998). Demolition and reuse of concrete and masonry: Demolition methods and practice. *Proceedings of the Second International Symposium.* Held by RILEM (the International Union of Testing and Research Laboratories for Materials and Structures), Nihon Daigaku Kaikan, Tokyo. November 7–11, 1988. Chapman and Hall, London.

Kawal, D.E. (1971). Information utilization in project planning. *Journal of the Construction Engineering Division*, ASCE, 97(2), 227–240.

Kein, A.T.T., Ofori, G., and Briffett, C. (1999). ISO 14000 series: Its relevance to the construction industry of Singapore and its potential as the next industry milestone. *Construction Management and Economics*, 17(4), 449–461.

Kemme, M.R. (1998). *Barcode Tracking System for Hazardous Waste* (HW). Construction Engineering Research Laboratory, U.S. Army Corps of Engineers, Champaign, IL, USA. <http://www.cecer.army.mil/td/tips/product/details.cfm?ID=25&TOP=1> (26 June 2005).

Kennedy, M. (2002). *The Global Positioning System and GIS: An Introduction.* Taylor & Francis, London.

Khalfan, M.M.A., Bouchlaghem, N.M., Anumba, C.J., and Carrillo, P.M. (2003). Knowledge management for sustainable construction: The C-SanD Project. In Molenaar, K.R. and Chinowsky, P.S. (Eds), *Winds of Change: Integration and Innovation in Construction (Proceedings of the 2003 Construction Research Congress)*, Honolulu, Hawaii, 19–21 March 2003. Reston: American Society of Civil Engineers.

Khasnabis, S., Alsaidi, E., Liu, L., and Ellis, R.D. (2002). Comparative study of two techniques of transit performance assessment. *Journal of Transportation Engineering*, ASCE, 128(6), 499–508.

Kincaid, J.E., Walker, C., and Flynn, G. (1995). *WasteSpec: Model Specifications for Construction Waste Reduction, Reuse, and Recycling.* Triangle J Council of Governments. P.O. Box 12276, Research Triangle Park NC 27709, USA. <http://www.tjcog.dst.nc.us/cdwaste.htm#wastespec> (26 June 2005).

Kloepfer, R.J. (1997). Will the real ISO 14001 please stand up. *Civil Engineering*, ASCE, 67(11), 45–47.

Koehn, E. (1976). Social and environmental costs in construction. *Journal of the Construction Division*, ASCE, 102(4), 593–597.

Lais, S. (1999). *Building Industry Braces for IT*, Online Onslaught, Computerworld Inc. <http://www.computerworld.com/cwi/story/frame/0,1213,NAV47_STO36776,00.html>.

Landscape Institute with IEMA (Institute of Environmental Management and Assessment) (2002). *Guidelines for Landscape and Visual Impact Assessment* (2nd edition). Spon Press, New York.

Laufer, A., and Jenkins, G.D. (1982). Motivating construction workers. *Journal of the Construction Division*, ASCE, 108(4), 531–545.

Launen, K.J. (1993). GPS – rapid solutions for transportation management, *Journal of Surveying Engineering*, ASCE, 119(1), 40–49.

Lavers, A.P., and Shiers, D.E. (2000). Construction law and environmental harm: The liability interface. *Construction Management and Economics*, Spon Press, 18(8), 893–902.

Leu, S., and Yang, C. (1999). GA-based multicriteria optimal model for construction scheduling. *Journal of Construction Engineering and Management*, ASCE, 125(6), 420–427.

Leu, S.S., Chen, A.T., and Yang, C.H. (1999). Fuzzy optimal model for resource-constrained construction scheduling. *Journal of Computing in Civil Engineering*, ASCE, 13(3), 207–216.

Leung, W.O. (1999). Effluent control for the construction projects. *Proceedings of International Conference on Urban Pollution Control Technology*, The Hong Kong Polytechnic University, Hong Kong, 449–455.

Li, H., and Love, P.E.D. (1997). Using improved genetic algorithms to facilitate time-cost optimization. *Journal of Construction Engineering and Management*, ASCE, 123(3), 233–237.

Li, H., Cao, J.N., and Love, P.E.D. (1999). Using machine learning and GA to solve time-cost trade-off problems. *Journal of Construction Engineering and Management*. ASCE, 125(5), 347–353.

Li, H., Chen, Z., and Wong, C.T.C., and Love, P.E.D. (2002). A quantitative approach to construction pollution control based on resource leveling. *International Journal of Construction Innovation*, 2(2), 71–81.

Li, H., Kong, C.W., Pang, Y.C., Shi, W.Z., and Yu, L. (2003a). Internet-based geographical information systems for e-commerce application in construction material procurement, *Journal of Construction Engineering and Management*, ASCE, 129(6), 689–697.

Li, H., Chen, Z., and Wong, C.T.C. (2003b). Application of barcode technology for an incentive reward program to reduce construction wastes in Hong Kong. *Computer-Aided Civil and Infrastructure Engineering*, 18(4), 313–324.

Lippiatt, B.C. (1999). Selecting cost-effective green building products: BEES approach. *Journal of Construction Engineering and Management*, ASCE, 125(6), 448–455.

Liska, R.W., and Snell, B. (1993). Financial incentive programs for average-size construction firm. *Journal of Construction Engineering and Management*, ASCE, 118(4), 667–676.

Lo, C.H. (2001). *Critical Factors for the Implementation of ISO 14001 Environmental Management System in Hong Kong Construction Industry*. M.Sc. Thesis. The Hong Kong Polytechnic University, Hong Kong.

Love, P.E.D., and Li, H. (2000). Quantifying the causes and cost of rework. *Construction Management and Economics*, 18(4), 479–490.

Lundberg, E.J., and Beliveau, Y.J. (1989). Automated lay-down yard control system-ALYC. *Journal of Construction Engineering and Management*, ASCE, 115(4), 535–544.

Maitra, A. (1999). Designers under CDM: A discussion with case studies. *Proceedings of the Institution of Civil Engineers: Civil Engineering*, 132(5), 77–84.

Mark, R. (2000). *IDC Online Previews New Portal*. Jupitermedia Corporation. <http://dc.internet.com/news/article.php/2101_389961> (26 June 2005).

Maslow, A.H., Stephens, D.C., and Heil, G. (1998). *Maslow on Management*. John Wiley, New York.

Masters, N. (2001). *Sustainable Use of New and Recycled Materials in Coastal and Fluvial Construction: A Guidance Manual*, Thomas Telford, London.

McCullouch, B.G., and Lueprasert, K. (1994). 2D bar-code applications in construction. *Journal of Construction Engineering and Management*, ASCE, 120(4), 739–753.

McCullough, C.A., and Nicklen, R.R. (1971). Control of water pollution during dam construction. *Journal of the Sanitary Engineering Division*, ASCE, 97(1), 81–89.

McFall, K. (2004). Delivering on promises (2/16/2004 Issue), ENR, The McGraw-Hill Companies, Inc. <http://enr.construction.com/features/technologyEconst/archives/040216.asp> (26 June 2005).

McMullan, R. (1998). *Environmental Science in Building* (4th edition). Macmillan, Basingstoke, England.

Meade, L.M., and Sarkis, J. (1999). Analyzing organizational project alternatives for agile manufacturing processes. *International Journal of Production Research*, Institution of Production Engineers, London, 37(2), 241–261.

Merchant, K.A. (1997). *Modern Management Control Systems: Text and Cases*. Prentice-Hall, Inc., New Jersey, USA.

Metcalf, D.D., and Urban, M.R. (1992). Leveraging the use of geographic information systems in highway corridor studies. In B.J. Goodno, J.R. Wright (Editors), *Computing in Civil Engineering and Geographic Information Systems Symposium* (Proceedings of the Eighth Technical Council on Computer Practices Conference held in conjunction with A/E/C Systems '92), ASCE, New York, 174–181.

Metro (1997). Construction site recycling-save money by recycling. <http:// www.metro-waste.com/commercial/construction.htm>, <http://www. multnomah.lib.or.us/metro/rem/rwp/constrcy.html> (2000/2003).

MFE (2004). Construction and demolition waste. Ministry for the Environment (MFE), New Zealand. <http://www.mfe.govt.nz/issues/waste/construction-demo/> (26 June 2005).

Middleton, F.M. and Stenburg, R.L. (1972). Research needs for advanced waste treatment. *Journal of the Sanitary Engineering Division*, 98(3), 515–528.

Mifkovic, C.S., and Peterson, M.S. (1975). Environmental aspects: Sacramento bank protection. *Journal of the Hydraulics Division*, ASCE, 101(5), 543–555.

Miller, R. (1999). *Demolition*. <http://www.fbe.unsw.edu.au/subjects/bldg/3005/demolition/ index.htm> (1 February 2001).

Mingpao.com (28 May 2002). *Transporters Allowed Free Disposal*, Mingpao, Hong Kong. <http://www.mingpao.com> (2 August 2002).

MOC (2001a). Announcement of First Authorized Qualifications of Main Contractors or Specialized Contractors with Special grade or Grade 1. Ministry of Construction (MOC), China. <http://zj.civil-engrg.com/ fg/show.asp?id=69> (22 March 2001).

MOC (2001b). Prescription for Qualification Management of Construction Enterprises. Prescription No. 87 of MOC. Beijing. Ministry of Construction (MOC), China. <http://www.cein.gov.cn/ad/d_1.htm,> <http://www.cin.gov.cn/indus/notice/2001072701.htm> and <http://www.cein.gov.cn/show.asp?rec_no=2590> (22 March 2001).

MOC (2001c). *Standard Grade of Main Building Contractors*, Ministry of Construction (MOC), China. <http://www.cein.gov.cn/ zznj/stepbystep/govfile/sgzcb.html> (22 March 2001).

Mohr, A.W. (1975). Energy and pollution concerns in Dredging. *Journal of the Waterways Harbors and Coastal Engineering Division*, ASCE, 101(4), 405–417.

Morris, D. (1976). Seasonal effects on building construction. *Journal of the Construction Division*, ASCE, 102(1), 29–39.

Morris, S.C., and Novak, E.W. (1976). Environmental health impact assessment. *Journal of the Environmental Engineering Division*, ASCE, 102(3), 549–554.

Mueller, I. *et al.* (1975). Waste exchange as a solution to industrial waste problems. *Israel Journal of Chemistry (ISJCAT Journal)*, 14, 226–233.

MWA (2000). *Commercial Programs – Construction and Demolition*, Metro Waste Authority (MWA), Des Moines, Iowa, USA. <http://www.metro-waste.com/commercial/construction.htm>, <http://www.multnomah.lib.or.us/metro/rem/rwp/ constrcy.html>.

NAHB Research Center (1999). *Guide to Developing Green Builder Programs*, A Report by the NAHB Research Center for the U.S. Environmental Protection Agency. <http://www.smartgrowth.org/pdf/Greengd.pdf> (12 June 2002).

Naresh, A.L., and Jahren, C.T. (1997). Communications and Tracking for Construction Vehicles, *Journal of Construction Engineering and Management*, ASCE, 123(3), 261–268.

Nasland, D.K., and Johnson, D.P. (1996). Real-time construction staking, *Civil Engineering*, ASCE, 66(6), 46–49.

Nelson, B. (1994). *1001 Ways to Reward Employees*, Workman, New York.

NOIE (2001). *B2B E-Commerce: Capturing Value Online*. National Office for the Information Economy (NOIE), Australia. <http://www.noie.gov.au/publications/NOIE/B2B/index.htm> (20 August 2002).

Norusis, M.J. (2000). *SPSS 10.0 Guide to Data Analysis*, Prentice-Hall, Inc., USA.

NPPA (1993). *Noise Pollution and Protection Act*, The People's Republic of China. Governmental Document in Chinese.

Ofori, G., Briffett, C., Gang, G., and Ranasinghe, M. (2000). Impact of ISO 14000 series on construction enterprises in Singapore. *Construction Management and Economics*, 18(8), 935–947.

OGC (2004). *Implementing Plans: Checklist of Organisational Learning and Maturity*. OGC Successful Delivery Toolkit™, The Office of Government Commerce (OGC), UK. <http://www.ogc.gov.uk/sdtoolkit/reference/tools/ip_positioning.html> (26 June 2005).

Olomolaiye, P.O., Jayawardane, A.K.W., and Harris, F.C. (1998). *Construction Productivity Management*, The Chartered Institute of Building. Addison Wesley Longman Limited, UK.

Orofino, J.F. (editor) (1989). *Structural Materials: Proceedings of the Sessions Related to Structural Materials at Structures Congress '89*, ASCE, New York.

Osuagwu, L. (2002). TQM strategies in a developing economy: Empirical evidence from Nigerian companies. *Business Process Management Journal*, MCB UP Ltd. 8(2), 140–160.

Parker, T. (1998). *Total Cost Indicators: Operational Performance Indicators for Managing Environmental Efficiency*, IIIEE Dissertations, the International Institute for Industrial Environmental Economics at Lund University, Sweden.

Parker, D.G., and Stader, T.N. (1995). Use of GIS to predict erosion in construction. In W.H. Espey, P.G. Combs (Editors), *Water Resources Engineering* (Proceedings of the First International Conference held in San Antonio, TX, 14–18 August 1995), New York: ASCE, 839–843.

Petts, J. (1996). *Environmental Assessment: Good Practice*. Proceedings of the Construction Industry Environmental Forum Conference on Good Practice in Environmental Assessment, Publication SP126, CIRIA, London.

Peurifoy, R.L. (2002). *Construction Planning, Equipment, and Methods* (6th edition), McGraw-Hill, New York.

Peyton, T.L. (1977). Energy management in commercial buildings. *Journal of Professional Activities*, ASCE, 103(1), 31–35.

Pilcher, R. (1992). *Principles of Construction Management* (3rd edition), McGraw-Hill, UK.

PIPS (2001). *Scanners and Data Collection Terminals*. Product Identification and Processing System (PIPS) Inc., New York, USA. <http://www.pips.com/scanners.html> (May 15, 2002).

PlanWare (2004). *Business Insight – Business Strategy Evaluator*, Invest-Tech Ltd. <http://www.planware.org/>

Poon, C.S., and Ng, L.H. (1999). The use of modern building technologies for waste minimization in Hong Kong. *Proceeding of International Conference on Urban Pollution Control Technology* (eds. C.S. Poon, X.Z. Li,). Hong Kong Polytechnic University, Hong Kong, 413–419.

Poon, C.S., Xu, Y., and Cheung, C.M. (1996). Building waste minimization in Hong Kong construction industry, *Journal of Solid Waste Technology and Management*, 23(2), 111–117.

Poon, C.S., Yu, A.T.W., and Ng, L.H. (2001). On-site sorting of construction and demolition waste in Hong Kong. *Resources, Conservation & Recycling*, 32(1), 157–172.

Proverbs, D.G., Holt, G.D., and Olomolaiye, P.O. (1998). A comparative evaluation of reinforcement fixing productivity rates amongst French, German and UK construction contractors. *Engineering, Construction and Architectural Management*, 5(4), 350–358.

Quazi, H.A., Khoo, Y.K., Tan, C.M., and Wong P.S. (2001). Motivation for ISO 14000 series Certification: Development of an evaluation model, *Omega: The International Journal of Management Science*. 29(6), 525–542.

Rappa, M. (2002). *Managing the Digital Enterprise*, Open Courseware. North Carolina State University. Raleigh, North Carolina, USA.

Rasdorf, W.J., and Herbert, M.J. (1989). Barcodes on the job site. *ID Systems*, 9(3), 32–36.

Rasdorf, W.J., and Herbert, M.J. (1990a). Automated identification systems-focus on bar coding. *Journal of Computing in Civil Engineering*, 4(3), 279–296.

Rasdorf, W.J., and Herbert, M.J. (1990b). Bar coding in construction engineering. *Journal of Construction Engineering and Management*, 116(2), 261–280.

Reardon, D.J. (1995). Turning down the power. *Civil Engineering*, ASCE, 65(8), 54–56.

Reddy, B.V.V., and Jagadish, K.S. (2003). Embodied energy of common and alternative building materials and technologies. *Energy and Buildings*, 35(2), 129–137.

Reeves, C.R. (1993). *Modern Heuristic Techniques for Combinatorial Problems*. Blackwell Scientific Publications, Oxford, Halsted Press, New York.

Rhatigan, J., and Irwin, D. (2001). *Salvage Style: 45 Home and Garden Projects using Reclaimed Architectural Details*, Lark Books, New York, N.Y. USA.

Robinson, S. (1994). *Successful Simulation: A Practical Approach to Simulation Projects*, McGraw-Hill, Maidenhead, England.

Robinson, G.L., Greening, W.J.T., DeKrom, P.W., Chrzanowdki, A., Silver, E.C., Allen, G.C., and Falk, M. (1995). Surface and underground geodetic control for superconducting super collider. *Journal of Surveying Engineering*, ASCE, 121(1), 13–34.

Rollett, H. (2003). *Knowledge Management: Processes and Technologies*, Kluwer Academic Publishers, Boston, USA.

Rosenfeld, Y., and Shapira, A. (1998). Automation of existing tower cranes: Economic and technological feasibility. *Automation in Construction*, Elsevier Science Inc., 7(4), 285–298.

Rosowsky, D.V. (2002). Reliability-based seismic design of wood shear walls. *Journal of Structural Engineering*, ASCE, 128(11), 1439–1453.

Ross, S., and Evans, D. (2003). The environmental effect of reusing and recycling a plastic-based packaging system. *Journal of Cleaner Production*, 11(5), 561–571.

Rutherford, J. (1981). On-site stormwater detention ponds: Wet and dry. In *Surface Water Impoundments: Proceedings of Symposium on Surface Water Impoundments* (1980: Minneapolis, Minn.). Edited by H.G. Stefan, ASCE, 972–984.

Saaty, T.L. (1996). *Decision Making with Dependence and Feedback: The Analytic Network Process*, Pittsburgh, PA, USA: RWS Publications.

Saaty, T.L. (1999). Fundamentals of the ANP. *ISAHP Proceedings*, Kobe, Japan. 16.

Saaty, T.L. (2001). Decision-making with the AHP: Why is the principal eigenvector necessary. *Proceedings of ISAHP 2001*, Berne, Switzerland.

Sacks, R., Navon, R., and Goldschmidt, E. (2003). Building project model support for automated labor monitoring. *Journal of Computing in Civil Engineering*, ASCE, 17(1), 19–27.

Sailor, V.L. (1974). Conservation of energy in buildings. *Journal of the Construction Division*, ASCE, 100(3), 295–302.

Salomon, V.A.P., and Montevechi, J.A.B. (2001). A compilation of comparisons on the analytic hierarchy process and other multiple criteria decision making methods: Some cases developed in Brazil. *Proceedings of 6th ISAHP*, Berne, Switzerland, 2–4 August 2001, 413–419.

SAP INFO (2004). *SAP INFO glossary*. <http://www.sap.info/public/en/glossary.php4/displayglossarystart> (18 November 2004).

Sauni, R., Oksa, P., Vattulainen, K., Uitti, J., Palmroos, P., and Roto, P. (2001). The effects of asthma on the quality of life and employment of construction workers. *Occupational Medicine*, Oxford University Press, 51(3), 163–167.

Sawhney, A., Mund, A., and Syal, M. (2002). Energy-efficiency strategies for construction of five star plus homes. *Practice Periodical on Structural Design and Construction*, ASCE, 7(4), 174–181.

SC of China (1998). *Managerial Ordinance on Environmental Protection of Construction Project*, The State Council of People's Republic of China. <http://www.envir.online.sh.cn/law/const2.htm> (22 November 2001).

Schodek, D.L. (1976). Effect of building regulations on built environment. *Journal of Professional Activities*, ASCE, 102(3), 293–300.

Schuette, S.D., and Liska, R.W. (1994). *Building Construction Estimating*. McGraw-Hill, Inc., New York.

Selwood, J.R., and Whiteside, P.G.D. (1992). Use of GIS for resource management in Hong Kong, In B.J. Goodno, and J.R. Wright (Editors), *Computing in Civil Engineering and Geographic Information Systems Symposium* (Proceedings of the Eighth Technical Council on Computer Practices Conference held in conjunction with A/E/C Systems '92), ASCE, New York, 942–949.

Senouci, A.B., and Adeli, H. (2001). Resource scheduling using neural dynamics model of Adeli and Park. *Journal of Construction Engineering and Management*, ASCE, 127(1), 28–34.

Seo, S., and Hwang, Y. (1999). An estimation of construction and demolition debris in Seoul, Korea: Waste amount, type, and estimation model. *Journal of the Air & Waste Management Association*, 49(8), 980–985.

Shorrock, L., *et al.* (1993). *Environmental Issues in Construction – A Review of Issues and Initiatives Relevant to the Building, Construction and Related Industries. Volume 1 – Overview Including Executive and Technical Summaries*. Publication SP93. CIRIA, UK.

Skibniewski, M.J., and Wooldridge, S.C. (1992). Robotic materials handling for automated building construction technology. *Automation in Construction*, 1(3), 251–266.

Skoyles, J.R. (1992). An approach to reducing materials waste on site. *The Practice of Site Management (Volume 4) (Ed.* Harlow, P.A.) The Chartered Institute of Building. Englemere, UK.

Skoyles, E.R., and Hussey, H.J. (1974). Wastage of materials. *Building*. 95–100 February.

Sparks, P.R., Liu, H., and Saffir, H.S. (1989). Wind damage to masonry buildings. *Journal of Aerospace Engineering*, ASCE, 2(4), 186–198.

Spivey, D.A. (1974a). Construction solid waste. *Journal of the Construction Division*, ASCE, 100(4), 501–506.

Spivey, D.A. (1974b). Environment and construction management engineers. *Journal of the Construction Division*, ASCE, 100(3), 395–401.

Srisoepardani, K.P. (1996). *The Possibility Theorem for Group Decision Making: The Analytic Hierarchy Process*, PhD Dissertation. Katz Graduate School of Business, University of Pittsburgh, Pittsburgh, Pennsylvania, USA.

Stanley-Miller Construction Company (1996). Case studies: ToolWatch® shines light on thieves: Stolen equipment recovered for Ohio Construction Company. <http://www.toolwatch.com/toolwatch/casestudies/Stanley.php> (15 May 2002).

Stone, P.A. (1983). *Building Economy* (3rd edition), Pergamon Press, England.

Stukhart, G. (1995). *Construction Materials Management*, Marcel Dekker, Inc., New York.

Stukhart, G., and Cook, E.L. (1989). *Barcode System Standardization in Industrial Construction*, Source Document 47, Construction Industry Institute. USA.

Stukhart, G., and Cook, E.L. (1990). Barcode standardization in industrial construction. *Journal of Construction Engineering and Management*, ASCE, 116(3), 416–431.

Stukhart, G., and Nomani, A. (1992). *Barcode System Standardization*, Source Document 70, Construction Industry Institute. USA.

Stukhart, G., and Pearce, S.L. (1988). Construction barcode standards. *Proceedings of fifth International Symposium in Robotics in Construction*, Japanese Society of Civil Engineers, 361–370.

Stukhart, G., and Pearce, S.L. (1989). Construction barcode standards. *Cost Engineering*, 31(6), 19–26.

Sukut (2003). *Unique Technologies Sukut has Implemented*. Sukut Construction, Inc., Santa Ana, USA. <http://www.sukut.com/# Anchor-Press-44097> (26 June 2005).

Suprenant, B.A. (Editor) (1990). *Serviceability and Durability of Construction Materials: Proceedings of the First Materials Engineering Congress*, ASCE, Denver, Colorado, USA.

Suprenant, B.A., and Malisch, W.R. (2000). The cost of waiting. *Concrete Construction*, Hanley-Wood, LLC, June 2000, 59–61.

Swain, J. (2001). Simulation software survey. *OR/MS Today*. February 2001 Issue. <http://www.lionhrtpub.com/orms/surveys/Simulation/Simulation.html> (26 June 2005).

Syswerda, G., and Palmucci, J. (1991). The application of genetic algorithms to resource scheduling. *Proceedings of the Fourth International Conference on Genetic Algorithms*. (Edited by R.K. Belew, and L.B. Booker), Morgan Kaufmann Publishers, San Mateo, California, USA. 502–507.

Tatum, C.B. (1978). Managing nuclear construction: an experience survey. *Journal of the Construction Division*, ASCE, 104(4), 487–501.

Taylor, D.C., Wilkinson, M.C., and Kellogg, J.C. (1976). A construction industry R&D incentives program. *Journal of Professional Activities*, ASCE, 102(3), 369–390.

Thomas, H.R., Sanders, S.R., and Bilal, S. (1992). Comparison of labor productivity. *Journal of Construction Engineering and Management*, ASCE, 118(4), 635–650.

Thomas, H.R., Sanvido, V.E., and Sanders, S.R. (1990). Impact of material management on productivity – a case study. *Journal of Construction Engineering and Management*, ASCE, 115(3), 370–384.

Tian, Q. (2002). *CMSS: An Interactive Construction Management Simulation System*, MASc Dissertation, Graduate Department of Civil Engineering, University of Toronto, Canada.

Tiwari, P. (2001). Energy tax and choice of house construction techniques in India. *Journal of Infrastructure Systems*, ASCE, 7(3), 107–115.

Trimble (2004). Mapping and GIS. <http://www.trimble.com/mgis.shtml> (30 June 2005).

Tse, R. (2001). The implementation of EMS in construction firms: Case study in Hong Kong. *Journal of Environmental Assessment Policy and Management (JEAPM)*, 3(2), 177–194.

Turban, E., King, D., Lee, J.K., and Viehland, D. (2003). *Electronic Commerce 2004: A Managerial Perspective* (3rd edition), Prentice Hall, USA.

Udo-Inyang, P.D., and Uzoije, C.H. (1997). HICIMS: An integrated GIS and DBMS application, In T.M. Adams (Ed.) *Computing in Civil Engineering* (Proceedings of the Fourth Congress held in conjuction with A/E/C Systems '97 in Philadelphia, PA, June 16–18, 1997), New York: ASCE, 240–247.

UNMFS (2004). Performance indicators for the evaluation of business plans, in *Multilateral Fund for the Implementation of the Montreal Protocol: Policies, Procedures, Guidelines and Criteria*. The Secretariat of the Multilateral Fund for the Implementation of the Montreal Protocol, Canada. <http://www.multilateralfund.org/policy.htm> (2004).

Uren, S., and Griffiths, E. (2000). *Environmental Management in Construction*, Publication C533, CIRIA, London, UK.

USEPA (1971). *Noise from Construction Equipment and Operations, Building Equipment, and Home Appliances*, Environmental Protection Agency, Washington, D.C., USA.

USEPA (1973). *Processes, Procedures, and Methods to Control Pollution Resulting from all Construction Activity*, Environmental Protection Agency, USA.

USEPA (2002a). EPA releases diesel exhaust health assessment. EPA Press Releases related to the Office of Transportation and Air Quality (OTAQ) for 2002. <http://www.epa.gov/otaq/press.htm> (30 September 2002).

USEPA (2002b). *Health Assessment Document for Diesel Engine Exhaust*. Prepared by the National Center for Environmental Assessment, Washington, DC, for the Office of Transportation and Air Quality; EPA/600/8-90/057F. Available from: National Technical Information Service, Springfield, VA; PB2002-107661. <http://www.epa.gov/ncea> (30 September 2002).

USEPA (2002c). Emission standards for new nonroad engines: Large industrial spark-ignition engines, recreational marine diesel engines, and recreational vehicles. Prepared by the National Center for Environmental Assessment, Washington, DC, for the Office of Transportation and Air Quality; EPA420-F-02-037. Available from: National Technical Information Service, Springfield, VA; PB2002-107661. <http://www.epa.gov/otaq/regs/nonroad/2002/f02037.pdf> (30 September 2002).

Vaid, K.N., and Tanna, A. (1997). *Wastage Control of Building Materials in Construction of Mass Housing Projects*, National Institute of Construction Management and Research (NICMAR), India, 1997, 19–35.

Valdez, H.E., and Chini, A.R. (2002). ISO 14000 standards and the US construction industry. *Environmental Practice*, 4(4), 210–219.

Varghese, K., and O'Connor, J.T. (1995). Routing large vehicles on industrial construction sites. *Journal of Construction Engineering and Management*, ASCE, 121(1), 1–12.

Velker, L. (1999). Caution: KM under construction – challenges and solutions from a geographically diverse contractor. *KMWorld Magazine*, 8(5).

Walter, C.E. (1976). Practical refuse recycling. *Journal of the Environmental Engineering Division*, ASCE, 102(1), 139–148.

Warren, F.H. (1973). Environmental impact on project schedules. *Journal of Professional Activities*, ASCE, 99(3), 299–306.

Warren, R.H. (1989). *Motivation and Productivity in the Construction Industry*, Van Nostrand Reinhold, New York.

Waugh, L.M., and Makar, J. (2001). *The Impact of e-commerce on the Construction Industry*. Presented at the international symposium Construction Innovation: Opportunities for Better Value and Profitability, June 6–7, 2001.

Wiegele, E. (2000). New construction methodologies using GIS, GPS, and the Web, *GITA's 9th Annual GIS for Oil & Gas Conference and Exhibition*, September 18–20–2000, in Houston, Texas, USA. <http://www.gisdevelopment.net/proceedings/gita/oil_gas2000/papers/wiegele.shtml> (30 June 2005).

Williams, J.W. (1992). Integrated GIS solutions with civil engineering projects, In B.J. Goodno, J.R. Wright (Editors), *Computing in Civil Engineering and Geographic Information Systems Symposium* (Proceedings of the Eighth Technical Council on Computer Practices Conference held in conjunction with A/E/C Systems '92), ASCE, New York, 328–331.

Wirt, D., Showalter, W.E., and Crouch, G. (1999). Integrating permanent equipment tracking with electronic operations and maintenance manuals. *Durability of Building Materials and Components 8 (Volume Four): Information Technology in Construction.* (Edited by M.A. Lacasse, D.J. Vanier) (CIB W78 Workshop) NRC Research Press, Ottawa, Canada, 2716–2723.

Wong, O., Morgan, R.W., and Kheifets, L. *et al.* (1985). Mortality among members of a heavy construction equipment operators union with potential exposure to diesel exhaust emissions. *British Journal of Industrial Medicine*, 42, 435–448.

WRC (2000). *Annual Review of Implementation of the Waste Reduction Framework Plan*, Waste Reduction Committee (WRC), Hong Kong. <http://www.info.gov.hk/wrc/>.

Zarli, A., Rezgui, Y., and Kazi, A.S. (2003). Present and future of European research on information technologies in construction. In Molenaar, K.R. and Chinowsky, P.S. (Eds), *Winds of Change: Integration and Innovation in Construction* (Proceedings of the 2003 Construction Research Congress, Honolulu, Hawaii, March 19–21, 2003). Reston: American Society of Civil Engineers.

Zeeger, C.V., and Rizenbergs, R.L. (1979). Priority programming for highway reconstruction. *Transportation Research Record*, No. 698, 15–23.

Zeng, S.X., Tam, C.M., Deng, Z.M., and Tam, Vivian W.Y. (2003). ISO 14000 and the construction industry: Survey in China. *Journal of Management in Engineering*, ASCE, 19(3), 107–115.

Zhu, Z. (1996). *Handbook of Building Construction Estimator*, China Construction Industry Press. Beijing, China.

Zutshi, A., and Sohal, A. (2004a). Environmental management system adoption by Australasian organisations: part 1: reasons, benefits and impediments. *Technovation*, 24(4), 335–357.

Zutshi, A., and Sohal, A. (2004b). A study of the environmental management system (EMS) adoption process within Australasian organizations: part 2: Role of stakeholders. *Technovation*, 24(5), 371–386.

Zyngier, S.M. (2002). Knowledge management strategies in Australia: Preliminary results of the survey of the knowledge management uptake in Australian companies. Caulfield East, Vic.: School of Information Management and Systems, Monash University, Australia.

## Collateral readings

BRE (1994). *Embodied Energy of Building Materials*. Building Research Establishment (BRE), UK. <http://www.ecosite.co.uk/depart/backinfo/bldmat.htm> (27 March 2003).

BRECSU (1996). *A Strategic Approach to Energy and Environmental Management*, Building Research Establishment. Garston, Watford, UK.

Bruhn-Tysk, S., and Eklund, M. (2002). Environmental impact assessment – a tool for sustainable development? A case study of biofuelled energy plants in Sweden. *Environmental Impact Assessment Review*, 22(2), 129–144.

Catterson, D. (1997). *The Reusable Building Materials Exchange: A New Approach to an Old Problem*. <http://www.olywa.net/speech/august97/catterson.html> (2000/2003).

CEIN (1998). *Main Indicators on Construction Enterprises*. <http://www.cein.gov.cn/home/hytj/jzyhytj/1998/14-8.doc> (22 November 2001).

CEIN (2001b). *The 10th Five-year Plan Opening Big Commercial Opportunity to Foreign Companies*. China Engineering Information Net (CEIN), China. <http://www.cein.gov.cn/show.asp?rec_no=577> (22 November 2001).

China EPB (1999). *Official Report on the State of the Environment in China 1998*. Environmental Protection Bureau (EPB), China. <http://www.zhb.gov.cn/bulletin/soechina98/index.htm> (22 November 2001).

China EPB (2001a). *Official Report on the State of the Environment in China 2000*. Environmental Protection Bureau (EPB), China. <http://www.zhb.gov.cn/bulletin/soe2000/index.htm> (10 June 2002).

China EPB (2001b). An Assessment Report on Integrated Environmental Governance in Forty-six Main Cities in China. Official Document of China EPB, No. (2001) 84. Environmental Protection Bureau (EPB), China. <http://www.zhb.gov.cn/sepa/news/200108/ hb84.htm> (5 June 2002).

China EPB (2001c, November 22). Conditions on Establishment of National Demonstration District based on ISO 14000 series. *The China Environment Daily*. Environmental Protection Bureau (EPB), China. <http://www.envir.online.sh.cn/info/2001/11/1122041.htm> (22 November 2001).

China NBS (1999). *Statistical Yearbook of China 1998*. National Bureau of Statistics (NBS), China. <http://www.stats.gov.cn/sjjw/ndsj/information/nj98/ml98.htm> (22 November 2001).

China NBS (2002). *China Statistic Yearbook (1995–2000)*. National Bureau of Statistics (NBS), China. <http://www.stats.gov.cn/tjsj/ndsj/index.htm> (25 June 2002).

CIIC (2001). *China in Brief: Administrative Division*. Internet Information Center, China. <http://www.china.org.cn/e-china/administrative/cities.htm> (22 November 2001).

CIRIA (2002). *Internet Scheme to Save on Waste*. CIRIA News. Issue 2, 8.

Clemen, R.T., and Reilly, T. (2001). *Making Hard Decisions with Decision Tools®.* Duxbury, Thomson Learning, USA.

CNPlus.co.uk (2002). *E-Construction Yearbook: Internet Services and Software for the UK Construction Industry.* EMAP. <http://www.cnplus.co.uk/econstruction/?ChannelID=62> (26 June 2005).

CSD (1984/2003). *Report on Annual Survey of Building, Construction and Real Estate Sectors.* Census and Statistics Department (CSD), Hong Kong.

ERM-Hong Kong, Ltd (1996). *Waste Reduction Study: Consultants' Findings and Recommendations.* EPD Waste Reduction Study (C1421). Environmental Protection Department (EPD), Hong Kong.

Green, S.B., Salkind, N.J., and Akey, T.M. (2000). *Using SPSS for Windows: Analyzing and Understanding Data* (2nd edition). Prentice-Hall Inc., Upper Saddle River, NJ, USA.

Horton Engineering (2001). *Horton Engineering Internet Calculators: Geometric Mean Calculator.* <http://www.graftacs.com/geohelp.html> (22 November 2001).

HPB (2000). *A Memorandum about Onerous Service on Disposal of Construction Waste.* Hefei Price Bureau (HPB), Anhui Province, China. <http://www.hfsr.gov.cn/fg5.htm> (22 November 2001).

Huang, Y.P. (2001). Fifteen hills of construction wastes on road. *Xian Evening Daily.* <http://cn.news.yahoo.com/ 011117/15/t3xc.html> (17 November 2001).

JACIC (2002). Construction by-products information exchange system and service. *JACIC News.* No. 156. May 2002. Japan Construction Information Center, Japan. <http://www.jacic.or.jp/books/jacicnews/jn156.pdf> and <http://www.recycle.jacic.or.jp> (26 June 2005).

JETRO (2000). Japanese Firms in China Foresee Better Sales This Year. Japan External Trade Organization (JETRO), Japan. <http://www.jetro.go.jp/it/e/press/2000/may2.html> (22 November 2001).

Jiang, Y. (1999). *EIA in China with Particular Reference to Suzhou Industrial Park.* MSc Thesis. School of Biological Sciences. University of Manchester. Retrieved from EIA Centre Reference Database. <http://quercus.art.man.ac.uk/eia/dbsrch/search_details.cfm?search_Article_ ID=6923> (30 July 2002).

Lu, A.Y., and Wu, M.C. (1999). Economic value of climate prediction information – the case of Taiwan construction Industry. *Workshop on the Impacts of the 1997/99 ENSO (El Niño/Southern Oscillation).* 1450–1510. <http://iri.columbia.edu/outreach/meeting/TWWS1999/> (22 November 2001).

MLIT (2001). *Construction Statistics Guidebook.* Ministry of Land, Infrastructure and Transport (MLIT), Japan. <http://www.mlit.go.jp/toukeijouhou/chojou/csg/csg_f.htm> (22 November 2001).

Ri, A. (1999, March 29). Environmental protection pipeline: A new transmit of construction waste on sites. *Construction Times.* <http://www.envir.online.sh.cn/info/np/file.asp?file=993-29-18.txt> (22 November 2001).

Saaty, T.L., and Vargas, L.G. (2001). *Models, Methods, Concepts and Applications of the Analytic Hierarchy Process,* International series in operations research and management science, Kluwer Academic, London.

Sawyer, T. (14 August 2000). Dot-coms need to court subcontractors. Document No.: MG20000823010000560. *Engineering News-Record.* 25(7), 35. <http://library.northernlight.com/MG20000823010000560.html?cb=7&dx=1005&sc= 0#doc> (15 January 2003).

SCH (2001). *Innovation in the Australian Building and Construction Industry, Year Ending December 2000*, Statistical Clearing House (SCH), Commonwealth Government of Australia. <http://www.sch.abs.gov.au> (22 November 2001).

SHAZAM (2000). *An Online Guide to SHAZAM: PART I – Learning the Basics: Statistics: Calculating a Geometric Mean.* <http://shazam.econ.ubc.ca/intro/gmean.htm> (26 June 2005).

Sina.com (2002). *China Quality Certification Center Put Family in Order.* <http://finance.sina.com.cn/b/ 20020719/234354.html> (23 July 2002).

Stenstadvold, M. (2000). The role of EIA in the planning and decision processes of large development projects in the Nordic countries: The case of the Gardermoen project. In Hilding-Rydevik, T. Editor, *EIA, Large Development Projects and Decision-making in the Nordic countries*, Nordregio Report 2001:6, ISBN 91-89332-18-0. 7-54. <http://www.nordregio.se/r0106.htm>, <http://www.nordregio.se/pdf/stenstadoh.pdf> (18 November 2004).

Tilford, K.R., Jaselskis, E.J., and Smith, G.R. (2000). Impact of environmental contamination on construction projects. *Journal of Construction Engineering and Management*, ASCE, 126(1), 45–51.

Transformation Strategies (1999). ISO 14000 mini-gap-analysis. <http://www.trst.com/gapanal.htm> (2 August 2002).

Tung, C.H. (1998). *Waste Reduction Framework Plan 1998–2007*, Government of Hong Kong SAR. Hong Kong.

Yang, S., Wen, T.Y., Wu, T.S., and Lee, Y.Y. (1989). *Research Method in Social and Behavioral Science* (13th edition), Tung-Hwa Book, Taiwan.

# Author index

# Subject index